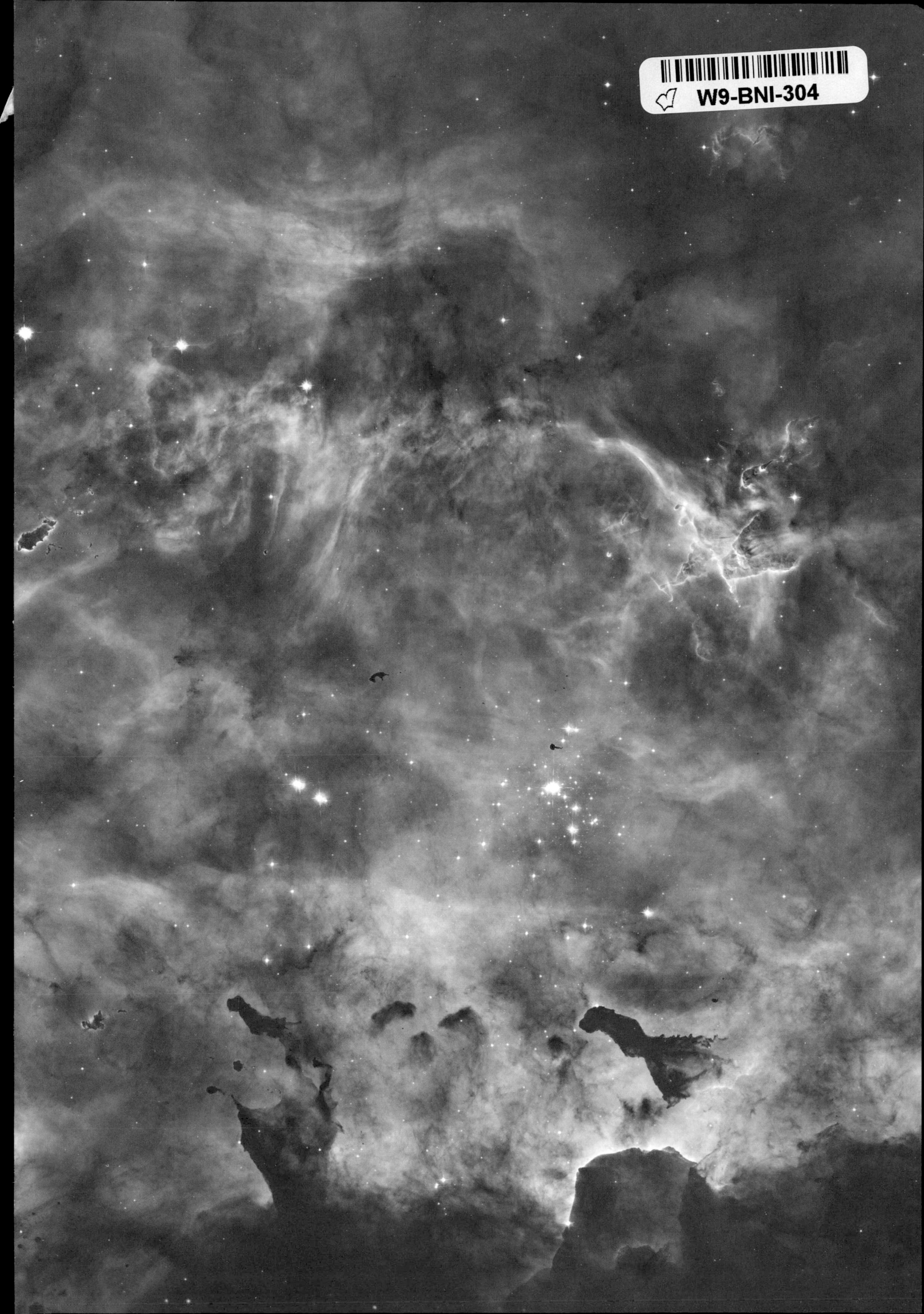

FIREFLY

SECOND EDITION

NIGHT SKY ATLAS

THE MOON, PLANETS, STARS AND DEEP SKY OBJECTS

ROBIN SCAGELL
WITH MAPS BY WIL TIRION

FIREFLY BOOKS

Published by Firefly Books Ltd. 2012

First printing

Publisher Cataloging-in-Publication Data (U.S.)

Scagell, Robin.
Night sky atlas : the moon, planets, stars and deep sky objects / Robin Scagell ; Wil Tirion.
2nd ed.
[128] p. : ill. (chiefly col.), maps, charts ; cm.
Includes index.
Summary: Maps and text show stars down to magnitude 5.5—all visible with binoculars or a small telescope. Opposite each map is a photo-realistic image that shows how the same portion of sky looks to the naked eye, allowing less experienced observers to quickly find specific objects of interest. The maps can be used for planning observations, navigating from one part of the sky to another, and for a quick reference guide.
ISBN-13: 978–1–7708–5142–9 (pbk.)
1. Astronomy — Charts, diagrams, etc. 2. Astronomy — Observer's manuals.
3. Stars — Charts, diagrams, etc. 4. Solar system — Maps. I. Tirion. Wil. II. Title.
523.8/0223 dc23 QB65.S335 2012

Library and Archives Canada Cataloguing in Publication

A CIP record for this title is available from Library and Archives Canada

Published in the United States by
Firefly Books (U.S.) Inc.
P.O. Box 1338, Ellicott Station
Buffalo, New York 14205

Published in Canada by
Firefly Books Ltd.
66 Leek Crescent
Richmond Hill, Ontario L4B 1H1

This book was developed by Philip's
a division of Octopus Publishing Group Ltd,
Endeavour House, 189 Shaftesbury Avenue,
London WC2H 8JY

Printed in China

Front cover: *(left)* star map *(Wil Tirion/Philip's)*; *(right)* photo-realistic star map *(Philip's)*. In the foreground is a 130 mm reflector *(Celestron International)*. In the background is the Moon at First Quarter *(Thierry Legault/Galaxy)*.

Back cover: *(above left)* nebulosity around Antares and Rho Ophiuchi *(Michael Stecker/Galaxy)*; *(above right)* Saturn *(Damian Peach/Galaxy)*; *(below)* constellation of Orion *(Wil Tirion/Philip's)*. In the background are stars near the galaxy NGC 4945 *(ESO)*.

Front endpaper: High-resolution view of part of the Eta Carinae Nebula, photographed by the Hubble Space Telescope. The region known as the Keyhole is seen at left *(STScI/Galaxy)*.

Rear endpaper: The star-forming area 30 Doradus in the Large Magellanic Cloud, photographed using the Hubble Space Telescope and the 2.2 m ESO telescope at La Silla, Chile *(STScI/Galaxy)*.

INTRODUCTION

This star atlas is for anyone who wants to learn the night sky, anywhere in the world. You need no prior knowledge – just add your own enthusiasm. Using a combination of maps of different scales and methods of plotting, you can discover the appearance of the stars wherever and whenever you observe.

A special feature of this atlas is realistic views of the constellations that match as closely as possible what you actually see in the sky, with no labels or grid lines to clutter the page. Facing each one is a conventional map of the same area that you can use to identify the stars and constellations.

Having found your way among the stars, you will want to study the other objects – the Sun, Moon and planets, and the much more distant nebulae, star clusters, and galaxies. The most interesting constellations are described in detail, with illustrations that show the objects of interest in a variety of ways, from drawings that match accurately what you can see through a small telescope to images taken through large telescopes, including the Hubble Space Telescope. Notes give you the basics of observational methods and help you to find and observe the objects, whether you have the simplest telescope or an up-to-the-minute computerized model.

A special website, www.stargazing.org.uk, accompanies this atlas to provide regularly updated links to sites giving further information and planetary positions, so you can always keep up-to-date. With this information, the *Firefly Night Sky Atlas* can be your astronomical companion for years to come.

CONTENTS

CHAPTER 1

GETTING STARTED

Learning the sky is a challenge, but the good news is that it is actually quite easy once you get started. What's more, it is something that you can do at your own pace, when the opportunity presents itself, and all you need is your eyes and a clear night sky. With a little effort you can recognize the major constellations or star patterns, and from there you can, if you want, move to the fainter ones. Soon, the sky will be as familiar to you as your own neighborhood.

Even if you live in a town you can make a good start. These days almost everyone is affected by light pollution – the glare from streetlights and other artificial forms of lighting. In towns and cities this almost drowns out the stars, but even in the largest city you can still see some stars and pick out the main constellations under the right conditions. People often say that when they see a really dark, clear sky, there are so many stars that they can't find their way around. So there is something to be said for starting with a simpler sky.

Probably the biggest hurdle that people encounter when they start to learn the sky is picking out their first constellations from the maps. One major difference between learning the sky and learning your way around a town is that the streets stay put, but the stars are always on the move. You may have been shown some well-known pattern in the past, but where is it now? Quite possibly it is not in the sky right now, or so low down that it is out of sight. Different stars are visible depending on the time of night and time of year, or where you happen to be on the Earth's surface. Those stars you learned a couple of months ago one early evening at home will be quite different from those you see after a late-night party on your foreign vacation. So the first thing to do is to find out what stars are visible at your time and place, then get your bearings so that you can locate them in the sky.

Then comes the next problem – the maps are small, but the sky is huge. You have to get the hang of the scale of the maps compared with the sky, and interpret the brightnesses of the stars as they are represented by the map symbols. There may be two or three planets in the sky which are not marked on the maps because they move around the sky at varying speeds. After a while you can pick out the difference between a planet and a star, and even work out which planet is which just by its appearance to the naked eye.

Quick start guide

This chapter explains how to follow the movements of objects in the sky, and how to know which map to choose for your particular time, date, and place. But if you want to plunge in at the deep end, go to Chapter 3 and pick out whichever small map is closest to your particular needs. That will show you the main stars visible looking either north or south. From there, you can choose whichever larger scale atlas maps you need from pages 26 to 41. Each atlas map is shown in two ways – in a photo-realistic version that closely resembles the real sky, and, on the opposite page, in a more conventional version in which the stars and constellations are labeled. For a more detailed look at each constellation, with targets suitable for binoculars and telescopes, look at the constellation maps in Chapter 8. These are arranged in groups so that the constellations visible on any particular night will be close together.

We have deliberately not shown every one of the 88 constellations in detail. This is because many of them contain little of interest, and few people spend much time looking at them, just as conventional travel guides don't bother listing every building in an area but concentrate on the more interesting ones. You may be surprised that some constellations everyone has heard of, such as Libra, don't get a special mention. But the well-known names of the zodiac are simply those that are along the path of the Sun and planets through the sky. In ancient times, people believed that the locations of the planets in the sky could affect their lives here on Earth. That has turned out not to be the case, but some of the rather ordinary suburbs of the sky have gained unwarranted fame as a result.

Telescopes and binoculars

You will probably want to find some of the sights of the sky using binoculars or a telescope. Some are easy to find, while others require more effort. People are sometimes disappointed that even an expensive telescope doesn't show anything resembling the dramatic color photos that you see in books. The sad fact is that the human eye, while incredibly acute and sensitive, is not good at picking out color in faint objects, while modern imaging techniques can transform a hazy blur into a vivid spectacle.

In our constellation pages we have shown objects in a variety of ways, from drawings and photographs made using amateur equipment to images made using giant telescopes or the Hubble Space Telescope. People often ask "Are those colors real?" The answer is usually "Yes, but they are not the colors you will see with your eyes." Though you may not be able to see the Crab Nebula, say, in the same way that a giant telescope shows it,

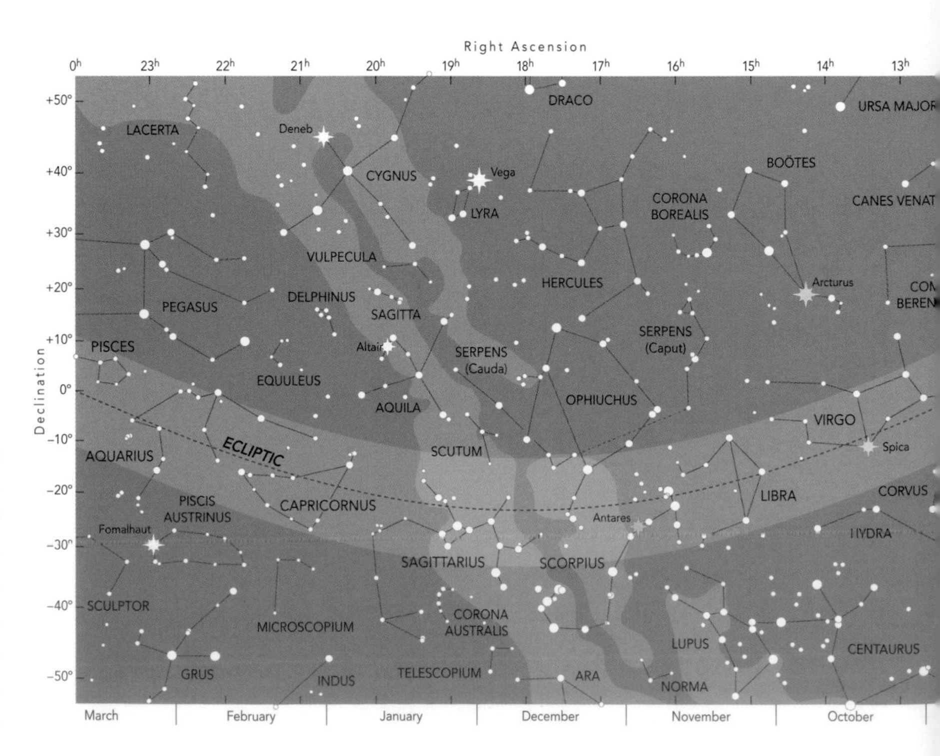

◄▲ *Two views of the Lagoon Nebula, M8, in Sagittarius. At left is an amateur photograph by Michael Stecker in California; above is a drawing by Darren Bushnall of Hartlepool, UK, using a 300 mm reflector. This telescope, though fairly large by amateur standards, shows much less detail than the photograph, and no color is visible.*

you can still get a sense of achievement from seeing that faint blur with your own eyes. The interest comes from tracking down an object rather than from the dramatic view.

Having said that, the sky does contain some more subtle gems. You may find a star cluster from your suburban lawn that looks like a faint sprinkling of stardust on the sky, or a pretty double star. The planets are visible even from light-polluted locations, while a sight of the Moon at First Quarter through even a small telescope can be stunning, exceeding the ability of any printed page to reproduce its appearance.

These days, many telescopes have what is called a Go To facility – that is, they are equipped with motors and a self-contained computer that will, if it is properly set up and everything works as it should, find hundreds of objects for you automatically. People often believe that this is the answer to the beginner's problem of finding their way around the sky. But many experienced stargazers would argue that it is better to learn to find objects for yourself, which will give you a much better knowledge of the sky than relying on technology. And knowing in advance what you are looking for is important, even with a well set-up Go To instrument. Are you looking for a tiny, faint fuzzy blob or a scattering of stars across the sky? The catalog number doesn't tell you. But our listings of objects tell you what to expect, and what sort of magnification to use when looking. And if you are relying on your own skill rather than electronics, we give many tips on how to find the objects from scratch.

For many people, visual observing is not enough. Some objects that are invisible to the naked eye can easily be imaged with either a digital camera or specialized CCD camera. Even in the city, some amazing results are possible using image processing that can be achieved on any modern computer. Details of methods used are on pages 20 and 21, and you can find the best targets for your camera in Chapter 8.

Whether you observe with the naked eye from the city, or with a computerized telescope from a dark-sky site, you will find plenty to keep you busy in the sky. We hope that this star atlas will be your companion for many years.

Getting your bearings

The first thing any stargazer needs to do is to establish which way is north or south. Most people have a good idea of their orientation from their home site, but if you are at a new observing site you may have problems. One way is of course to look at a map, while another is to notice the direction of the Sun at true midday. At this time it is on your *meridian* – your north–south line. This means that it is due south if you are in the northern hemisphere, or due north if you are in the southern. In the tropics it will be more or less overhead, which is not so useful. Wherever you are on Earth, the Sun and indeed the rest of the sky moves slowly

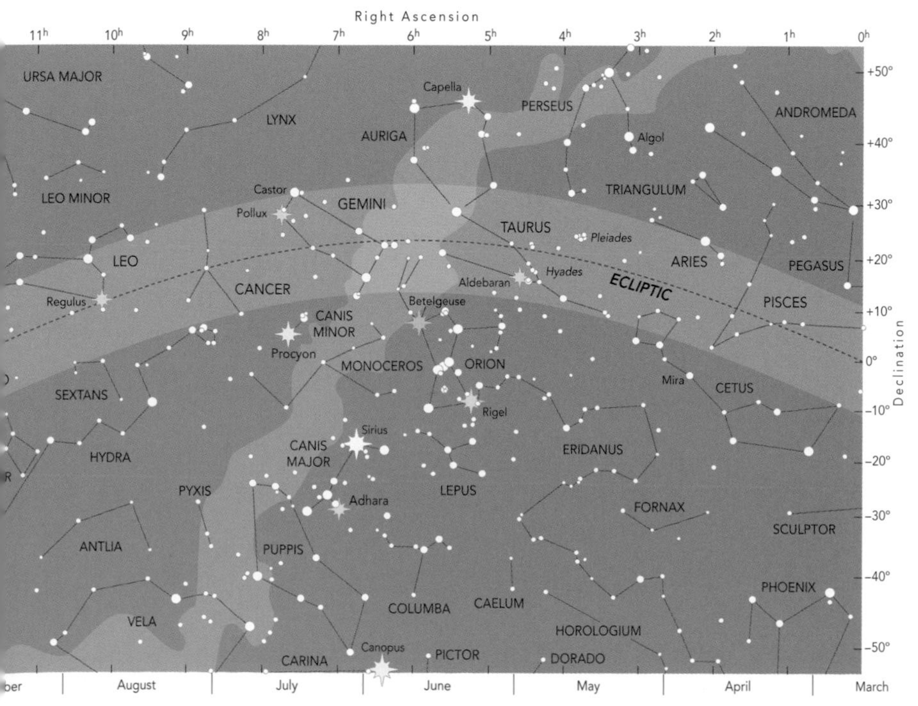

◄ *The apparent path of the Sun through the sky is called the ecliptic. The zodiacal constellations are those that lie along this path.*

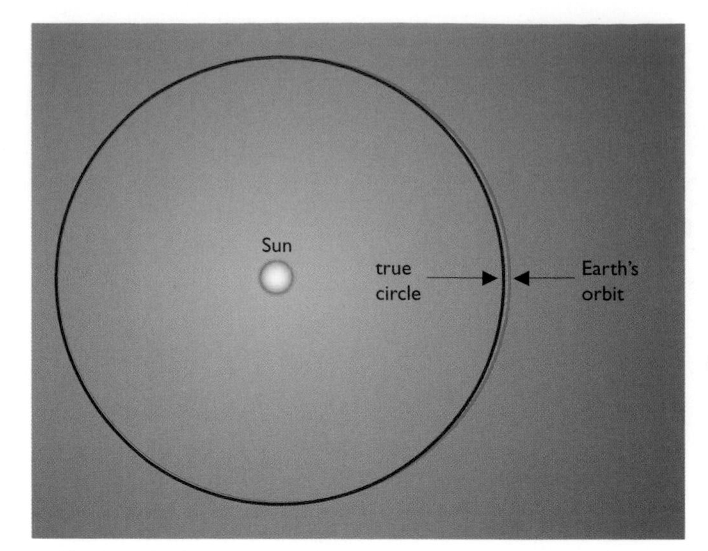

▲ *The Earth's elliptical orbit (blue) compared with a true circle (red). The difference appears slight, but the Earth's distance from the Sun varies by 5 million km.*

as the Earth turns, rising in the eastern side of the sky and setting in the western side. Only rarely does the Sun actually rise due east, and set due west.

In practice, other factors affect the actual position of the Sun at midday. The most obvious is the presence or absence of Summer or Daylight Saving Time. When this is in force midday usually occurs an hour later, around 1 pm. Another factor is an unevenness in the Sun's apparent motion over the year, which results in it being sometimes late and sometimes early to reach the meridian. This is caused by the Earth's orbit around the Sun not being circular but being an ellipse – a slightly squashed circle, with the Sun slightly displaced from the center. Virtually all orbits, whether of planets or comets around the Sun or one star in a double star around the other, are elliptical. When the Earth is closest to the Sun, it moves faster in its orbit than when it is most distant, resulting in the varying speed of the Sun through the sky. These daily variations are known as the Equation of Time, and you can find them listed on the website www.stargazing.org.uk. The same factors affect the accuracy of sundials.

A final correction in the Sun's position at midday arises from your position within your time zone on Earth. We all know that the time in different countries east or west of our own is different, but for convenience we all stick to the same time within our own time zone. If you happen to be near the edge of your zone, however, you could find that time as measured by the Sun is different from that on your watch by an additional half an hour or more. These factors also affect the local times of the appearance of the sky as shown on the star maps, so if the maps always seem to be slightly in error, this could be the reason.

The celestial sphere – how the sky moves

The changes in the Sun's position, and the overall movement of the sky, are most easily understood by thinking of the sky in terms of a huge sphere surrounding the Earth. We know that in reality the Earth goes round the Sun, and that it is the rotation of the Earth that causes the Sun, Moon, and stars to rise and set. But the concept of the celestial sphere is very useful, and is the standard way of looking at such things.

A simple celestial sphere is shown below. It is drawn so that the Sun's apparent path through the sky – called the *ecliptic* – is horizontal and the Earth's axis is tilted to it. Imagine yourself on the Earth in the middle; you would feel that the Earth is stationary and the giant sphere surrounding the Earth appears to turn once a day along the tilted axis that sticks through the Earth, from north to south pole.

Just like the Earth, the celestial sphere has a north and south pole, and an equator. These are the extensions into the sky of the same locations on the Earth. The sky also has two hemispheres, northern and southern.

You can see that if you were at the North Pole, the sky and stars would rotate parallel to the horizon, with none rising or setting. At the South Pole they would do the same thing, but in the opposite direction. Stand on the Equator and everything appears to rise and set vertically from the horizon, with objects toward the north and south performing smaller and smaller semicircles as they get closer to each celestial pole.

Most of us, however, live at some intermediate point on the Earth's surface. We see one of the celestial poles part of the way up the sky, with all the stars rotating around it. In the Earth's northern hemisphere you can see the stars around the north celestial pole all the time, and you can also see some of the stars in the sky's southern hemisphere, below the celestial equator. But there are some stars, around the south celestial pole, that you will never be able to see because the Earth is always in the way. The opposite applies at locations in Earth's southern hemisphere. People living at the Equator can see the whole sky at one time or another, while the nearer to one of the poles you live, the less of the sky's opposite hemisphere is visible to you.

Wherever you live, objects in the sky will always appear to rise along the eastern horizon and set along the western horizon. In the northern hemisphere they reach their highest point above the horizon toward the south, which is where you will find the Sun at true midday. In the southern hemisphere objects rise from the

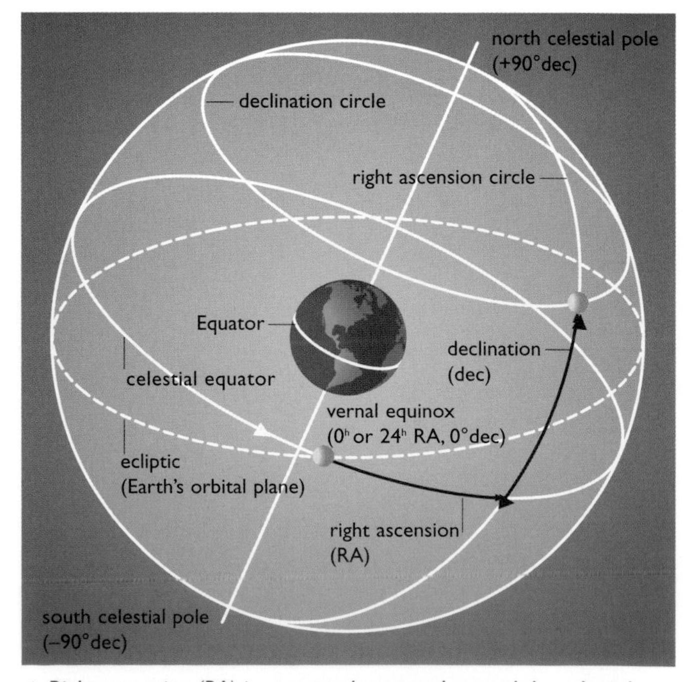

▲ *Right ascension (RA) is measured eastward around the celestial equator from the First Point of Aries. Declination (dec) is given by the angle north or south of the celestial equator.*

► A time exposure image of star trails taken at 44°N using a fisheye lens looking east. The Pole Star at top left remains almost stationary, while stars in the east rise at an angle to the horizon. Due south, at far right, the stars set only a short time after they have risen.

eastern horizon at the reverse angle, and reach their highest point due north.

Star trail photographs are among the easiest of astronomical photographs to take, and simply involve keeping the camera's shutter open for a period of time so that the stars trail across the sensor. Photos taken in this way show the stars' movements very clearly.

The Sun's movements

The Earth's axis is inclined at an angle – $23\frac{1}{2}°$ – to the ecliptic (which is in actual fact the plane of the Earth's orbit around the Sun). This is what gives rise to the seasons, and from the stargazer's point of view it means that the ecliptic is inclined to the celestial equator. The Sun moves from west to east along the ecliptic during the course of the year while the Earth turns daily beneath the stars. As we measure time by the Sun, it is the average interval between successive middays that we divide into 24 hours. But because the Sun has moved slightly along the ecliptic, there is a slight slippage between the time as told by the Sun, and that as told by the stars. In fact, the same stars are visible four minutes earlier each night.

After a year the Sun has moved full circle and all the four minutes have added up to get the stars back to almost exactly the same position as they were the previous year at that date. They are not in exactly the same place, which is why every four years, in what is known as a leap year, we need an extra day to keep the movements of the Sun and the stars together. Without this extra day, eventually the months and the Sun's movements would get out of step, and June or December would end up occurring in autumn or spring.

When the Sun is north of the celestial equator, it is high in the sky in the northern hemisphere. It reaches its maximum altitude around June 21, which is midsummer in the northern hemisphere. From there it moves southward until around September 22 it is exactly on the celestial equator. On this day only it rises exactly in the east, and day and night are 12 hours long all over the globe (excepting the regions around the poles where it is on the horizon). This is the *equinox*, meaning "equal nights." Three months later it is at its southernmost point, and the southern hemisphere experiences midsummer; then in March it is back again at the equator for the vernal equinox.

The ecliptic is also the approximate track of the Moon and planets, though they can move several degrees on either side of it. We are all familiar with the Sun – and therefore the ecliptic – being high in summer at midday, but look at the diagram and you will see that at midnight the opposite occurs, and the ecliptic is low in the sky. As a result, the Moon and planets are rather low in the sky at midsummer. Then in midwinter, the ecliptic is high in the sky at midnight and any planets that happen to be roughly opposite the Sun in the sky will appear very high up, which makes them ideally placed for observation.

Positions in the sky

The celestial sphere diagram also helps to explain the way positions in the sky are measured – the celestial equivalent of latitude and longitude. The easiest to follow is the equivalent of latitude, which is called *declination*. Just as on Earth, this is measured in degrees from the equator to the pole, with the equator being 0° and the pole being 90°. Southerly declinations are usually given a minus sign. If you are at the Equator, the stars directly overhead have a declination of 0°, and wherever you are on Earth, stars directly above you (in your *zenith*) have a declination equal to your latitude.

Longitude on Earth is measured in degrees west of a fixed point on the Earth's surface, which by international agreement is Greenwich Observatory near London. The equivalent fixed point in the sky is not a particular star, but the point where the ecliptic crosses the celestial equator when the Sun is on its way northward in March. This is known as the vernal equinox or First Point of Aries, though because of a slow movement of the Earth's axis, known as precession, it is no longer in Aries.

The celestial equivalent of longitude is known as *right ascension*, or RA. Though RA can be measured in degrees, it is usually measured in hours and minutes eastward from the First Point of Aries, from 0 hours to 24 hours. There is a good reason for this – it relates to the apparent rotation of the celestial sphere around the Earth. Say you start observing with the First Point of Aries on your meridian. An hour later, stars with an RA of 1 hour will be on your meridian. At that time, stars with an RA of about 7 hours and on the celestial equator will be just rising in the east. The sky is one great 24-hour clock, and the scale of RA reflects this.

Bear in mind that the separation of two stars 6 hours of RA apart depends on their declination. If you look toward the pole

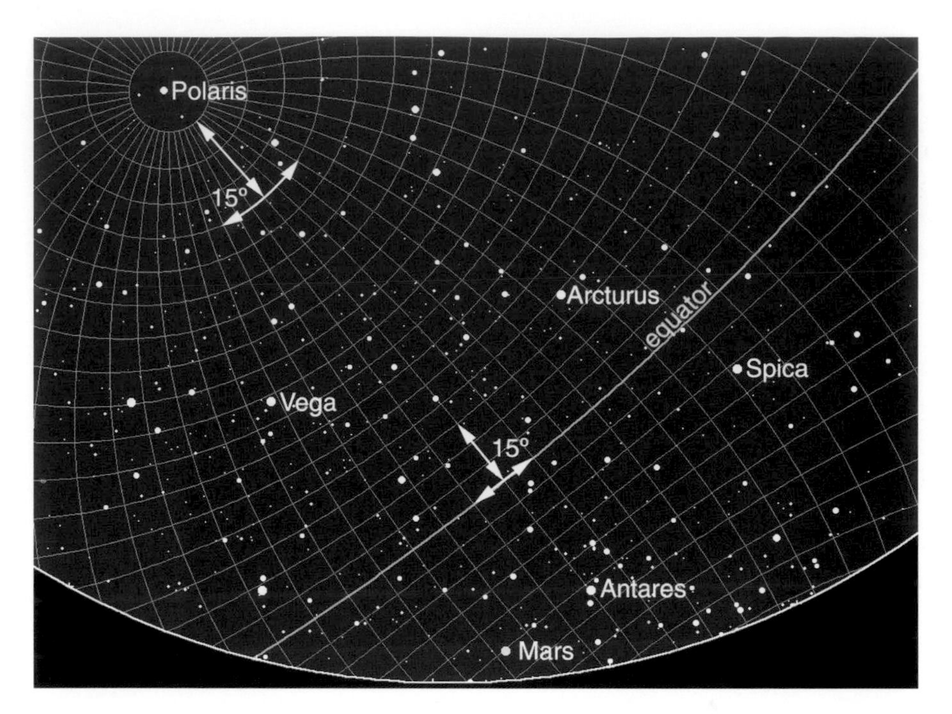

◄ The area of the image on page 7, with the RA and dec lines added. RA lines are marked every 30 minutes, and dec lines every 5°. At the celestial equator, 15° of dec is the same as 1 hour of RA. But near the pole, 15° of dec is the same as 3 hours of RA. The map scale changes near the pole to retain the appearance of the constellations.

where all the hour lines of RA converge, two stars separated by 6 hours of RA will actually be close together in the sky. On the celestial equator, however, they will be half a sky diameter apart.

Angles in the sky

How do these measurements relate to the separation of two stars in the sky? We usually measure angles in degrees, so two objects exactly on opposite sides of the sky are 180° apart, while the distance between the horizon and the zenith is 90° and an object halfway up the sky is 45° above the horizon. Stargazers often refer to such angles, and it is very useful to have a mental idea of the appearance of angles in the sky. If you hear that Mercury, for example, will be 10° above the horizon this evening, it is worth knowing how high this is in relation to the local rooftops. Or if the Moon will be 3° from a particular star, will that be within the field of view of your binoculars, or a telescope?

The degrees of declination are the same as degrees of angular measurement. So two stars that have the same RA but are separated by 22 degrees of declination, say, are a genuine 22° apart in the sky. An hour of RA is 15° only when it is on the celestial equator. Closer to the poles, an hour of RA will be a smaller angle, which is one reason why RA is usually referred to in hours and minutes rather than degrees.

It so happens that a handspan from tip of thumb to little finger, when stretched out at arm's length, is about 16° to 20°. People often use this to get an idea of scale when measuring angles in the sky. In the same way, your index finger at arm's length is about 1°. Other useful angles to remember are the separation of the two stars known as the Pointers in the Big Dipper or Plough (see page 84), which are 5.4° apart, or, in the southern hemisphere, the length of the Southern Cross, which is 6°. A typical field of view of binoculars is about 5°, and a telescope's low-power eyepiece is often about 1°.

From time to time you may find references to positions in the sky in terms of *altitude* and *azimuth*. In this case, altitude refers to the angular distance in degrees of an object up from the horizon. In the earlier example, therefore, Mercury would have an altitude of 10°. This is not to be confused with the altitude of your observing site above sea level! Azimuth is the angle round the horizon starting from north and going through east. North is 0°, east is 90°, south is 180°, and west is 270°. This applies in both the northern and the southern hemispheres. Each degree is divided into 60 arc minutes, which in turn are each divided into 60 arc seconds. Altitude and azimuth are most commonly used to refer to the positions of artificial satellites in the sky. The position where an object rises along the horizon is also often referred to by its azimuth, as this allows you to work out its direction on a terrestrial map.

Getting to know the sky

Learning the sky is little different from learning your way around any new district that you happen to be in. To begin with you need to know the overall structure with a few landmarks that you can easily recognize. Then you can start to fill in the details to suit your own requirements. In some ways, learning the sky is easier because you get an instant overview anyway, as if you were seeing the district from the air.

To begin with, you need to establish which stars are visible at the time you will be observing, from your location. The seasonal maps in Chapter 3 give you an overview of the sky looking either north or south, and from either the northern or the southern hemisphere. From these, you can refer to the appropriate atlas map for a more detailed view of that part of the sky.

Remember that the lower map of each pair shows the view looking toward the pole, in which case the only change from month to month is the orientation of most of the stars. These stars that never rise or set are known as *circumpolar stars*. But the upper map shows the sky looking away from the pole – south in the northern hemisphere and north in the southern hemisphere. In this direction, the stars on view change with the seasons.

The scale of these maps is a little small for picking out the constellations, so you may prefer to use the main atlas maps to which they refer instead. The key to finding your way here is to identify the brightest stars or most obvious patterns, as referred to in the captions to the seasonal maps. Be aware that any bright planets in your region of sky will make a big difference to the star patterns. You can check the monthly positions of the bright planets by using the tables on the website www.stargazing.org.uk.

As you observe, the stars move slowly across the sky. After an hour or so, some stars in the west are noticeably lower, while others in the east are higher up. If you watch for several hours, a completely new sky presents itself. Alternatively, if you observe at the same time each night, the same thing happens over a period of time. In a month of observing, you will see stars that you would have waited two hours to see on the first night of observation. Should you be impatient to see stars that will be in your evening sky in three months' time, you can wait for six hours and they will be there in the early morning sky. After a year, the sky has gone full circle and you are back where you started.

An alternative to the seasonal maps in Chapter 3 is to purchase a planisphere for your particular latitude. These handy star disks have a map of the sky visible from your location with a rotatable overlay that can be set to show exactly which stars will be above the horizon at any chosen date and time.

Star brightnesses

On the maps, the stars are shown by dots of different sizes according to their brightness. The true range of brightness of the stars is enormous – the brightest star in the night sky is 1000 times brighter than the faintest star easily visible with the naked eye. To make the area of its dot accurately depict its brightness compared with the faintest star on the map, it would have to be about 8 mm across, which would be impractical. This is why we have created the photo-realistic maps, in which the fainter stars are shown by darker dots. This, and the lack of labels, makes them much more representative of the real night sky. Once you have identified the star patterns using the photo-realistic maps, you can use the conventional maps to find their names.

Star brightnesses are measured on a scale unique to astronomy, which has remained in place since ancient Greek times. Put simply, the brightest stars are 1st magnitude while the faintest ones are 6th. Thus they are ranked like the winners in a contest, with the most prestigious having the lowest number.

EXAMPLES OF TYPICAL MAGNITUDES			
Sun	–26.8	Typical naked-eye limit in country	5.8
Moon	–12.7	Naked-eye limit with acute vision	7
Venus at its brightest	–4.4	10 × 50 binocular limit	11
Sirius (brightest star)	–1.5	Proxima Centauri (nearest star)	11.01
Alpha Centauri (nearby star)	–0.3	114 mm telescope limit	13
Polaris (Pole Star)	2.1	200 mm telescope limit	14
Typical naked-eye limit in towns	4.5	Giant telescope limit	30+

This is the opposite of most means of measurement, and the magnitude scale is a potential source of confusion at all levels in astronomy.

When it was required to place the ancient scale on a mathematical basis, it was decided to make five magnitudes exactly equal to a range of 100 times in brightness. To accommodate the full range of star brightnesses, the very brightest objects had to be assigned negative magnitudes. The table gives a few representative magnitudes. Very soon you will get an idea of what each magnitude means, and how a star of, say, the 2nd magnitude appears to the naked eye from your usual observing site, and when seen through binoculars or a telescope.

This is one of the great keys to finding objects in the sky, because you need to be able to know when you have found the star you are looking for, and by how much your optical aid brightens its appearance.

Constellations and their names

Beginners to astronomy are often bemused or even amused by the peculiar names of the constellations and their meanings. What is fishy about Pisces, and what is balanced about Libra? The star patterns themselves rarely work on a "join the dots" basis. But the constellations are actually quite useful, even if the names have little to do with their appearance. Say to stargazers that a particular object is in Aquarius, and they will immediately have an idea of which part of the sky you are talking about, and when it is visible. But say that it is on the celestial equator and at 22 hours RA, and most will have to turn to a star map to see where you mean.

Patterns are very useful to us when identifying things, and the fact is that the stars are distributed pretty much at random rather than in well-organized groups. Even if we were to give them modern names, as is sometimes suggested, they would not represent the new objects any better than they do a centaur or an eagle, for example.

The constellation names are derived from several sources. The very oldest predate history itself, and came from the Middle East. They have been added to over the years, in Greek, Roman, and medieval Arabic times, and there was a flurry of additions during the 18th century. In the early 20th century their borders were regularized. All cultures had their own names, often for different patterns of the same stars, but only these classical names are in use worldwide today. In English-speaking countries it is usual to apply the classical version (such as Leo) rather than its translation (the Lion).

▼ The constellation patterns rarely look like what they are supposed to represent. But the use of shapes makes it much easier to recognize the various parts of the sky. The lines have no official significance, and are shown on maps as an aid to picking out the patterns.

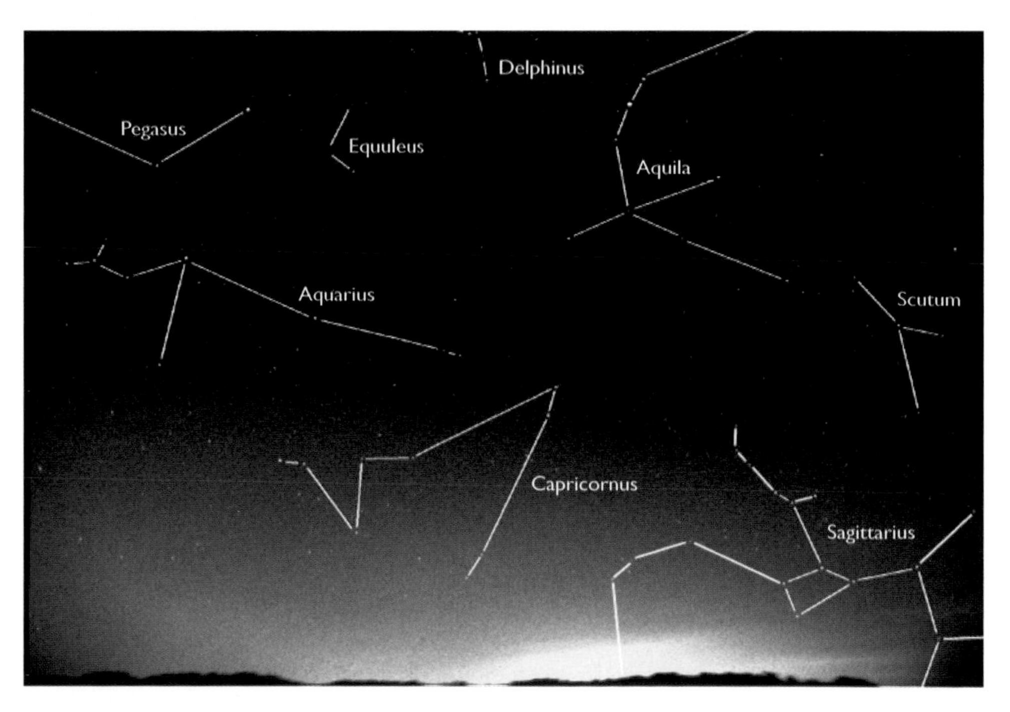

The sky therefore contains a motley collection of heroes, villains, creatures commonplace and fantastic, and a few items of hardware such as a lyre, a set of scales, and a telescope (one of the 18th-century introductions).

Of the 88 constellations now recognized, 48 were in use in Greek and Roman times, including most of the well-known ones. Most of these refer either to animals or to mythological characters. We can picture people sitting round fires in the open, telling tales and using the star patterns to represent the characters in the stories. Today we watch television instead, but we can think of the constellation heroes as being the soap stars of the ancient past.

These mythological characters often lent their figures to the names of the stars that comprise them. The name of the star Betelgeuse in Orion is a classic example. It is said to derive from the Arabic for "The Hand of Jauzah the Giant." Rigel is the same giant's leg or foot. Many of the names come from Arabic words, because the Arab world maintained the flame of knowledge during what are known as the Dark Ages in Europe. This is why many of them begin with "Al-," meaning "The."

In addition to the names, stars have other designations. In the 17th century the principal stars in each constellation were assigned Greek letters by German astronomer Johann Bayer. Usually the brightest star is alpha (α), the next brightest beta (β), and so on. The constellation name is turned into its Latin genitive form, as in Alpha Centauri, meaning "alpha of Centaurus," Delta Cephei ("delta of Cepheus"), and so on. So in addition to learning the constellations, the knowledgeable stargazer should know the Latin genitive version of its name as well.

There are only 24 Greek letters, and in the 18th century when more stars were accurately cataloged, it became necessary to assign numbers. This was first done by the British Astronomer Royal of the day, John Flamsteed, who numbered the stars in his catalog, again constellation by constellation, beginning at the westernmost edge. These Flamsteed numbers sit alongside the

THE CONSTELLATIONS

Name	Genitive	Abbreviation	Common name	Name	Genitive	Abbreviation	Common name
Andromeda	Andromedae	And	Andromeda	Leo	Leonis	Leo	Lion
Antlia	Antliae	Ant	Air Pump	Leo Minor	Leonis Minoris	LMi	Little Lion
Apus	Apodis	Aps	Bird of Paradise	Lepus	Leporis	Lep	Hare
Aquarius	Aquarii	Aqr	Water Bearer	Libra	Librae	Lib	Scales
Aquila	Aquilae	Aql	Eagle	Lupus	Lupi	Lup	Wolf
Ara	Arae	Ara	Altar	Lynx	Lyncis	Lyn	Lynx
Aries	Arietis	Ari	Ram	Lyra	Lyrae	Lyr	Lyre
Auriga	Aurigae	Aur	Charioteer	Mensa	Mensae	Men	Table (Mountain)
Boötes	Boötis	Boo	Herdsman	Microscopium	Microscopii	Mic	Microscope
Caelum	Caeli	Cae	Chisel	Monoceros	Monocerotis	Mon	Unicorn
Camelopardalis	Camelopardalis	Cam	Giraffe	Musca	Muscae	Mus	Fly
Cancer	Cancri	Cnc	Crab	Norma	Normae	Nor	Level (square)
Canes Venatici	Canum Venaticorum	CVn	Hunting Dogs	Octans	Octantis	Oct	Octant
Canis Major	Canis Majoris	CMa	Great Dog	Ophiuchus	Ophiuchi	Oph	Serpent Bearer
Canis Minor	Canis Minoris	CMi	Little Dog	Orion	Orionis	Ori	Orion
Capricornus	Capricorni	Cap	Sea Goat	Pavo	Pavonis	Pav	Peacock
Carina	Carinae	Car	Keel (of a ship)	Pegasus	Pegasi	Peg	Pegasus (winged horse)
Cassiopeia	Cassiopeiae	Cas	Cassiopeia	Perseus	Persei	Per	Perseus
Centaurus	Centauri	Cen	Centaur	Phoenix	Phoenicis	Phe	Phoenix
Cepheus	Cephei	Cep	Cepheus	Pictor	Pictoris	Pic	Easel
Cetus	Ceti	Cet	Whale	Pisces	Piscium	Psc	Fishes
Chamaeleon	Chamaeleontis	Cha	Chameleon	Piscis Austrinus	Piscis Austrini	PsA	Southern Fish
Circinus	Circini	Cir	Compass	Puppis	Puppis	Pup	Stern (of a ship)
Columba	Columbae	Col	Dove	Pyxis	Pyxidis	Pyx	Compass
Coma Berenices	Comae Berenices	Com	Berenice's Hair	Reticulum	Reticuli	Ret	Net
Corona Australis	Coronae Australis	CrA	Southern Crown	Sagitta	Sagittae	Sge	Arrow
Corona Borealis	Coronae Borealis	CrB	Northern Crown	Sagittarius	Sagittarii	Sgr	Archer
Corvus	Corvi	Crv	Crow	Scorpius	Scorpii	Sco	Scorpion
Crater	Crateris	Crt	Cup	Sculptor	Sculptoris	Scl	Sculptor
Crux	Crucis	Cru	Southern Cross	Scutum	Scuti	Sct	Shield
Cygnus	Cygni	Cyg	Swan	Serpens	Serpentis	Ser	Serpent
Delphinus	Delphini	Del	Dolphin	Serpens Caput			Serpent's head
Dorado	Doradus	Dor	Goldfish or Swordfish	Serpens Cauda			Serpent's tail
Draco	Draconis	Dra	Dragon	Sextans	Sextantis	Sex	Sextant
Equuleus	Equulei	Equ	Foal	Taurus	Tauri	Tau	Bull
Eridanus	Eridani	Eri	River Eridanus	Telescopium	Telescopii	Tel	Telescope
Fornax	Fornacis	For	Furnace	Triangulum	Trianguli	Tri	Triangle
Gemini	Geminorum	Gem	Twins	Triangulum Australe	Trianguli Australis	TrA	Southern Triangle
Grus	Gruis	Gru	Crane	Tucana	Tucanae	Tuc	Toucan
Hercules	Herculis	Her	Hercules	Ursa Major	Ursae Majoris	UMa	Great Bear
Horologium	Horologii	Hor	Pendulum Clock	Ursa Minor	Ursae Minoris	UMi	Little Bear
Hydra	Hydrae	Hya	Water Snake	Vela	Velorum	Vel	Sails (of a ship)
Hydrus	Hydri	Hyi	Lesser Water Snake	Virgo	Virginis	Vir	Virgin
Indus	Indi	Ind	Indian	Volans	Volantis	Vol	Flying Fish
Lacerta	Lacertae	Lac	Lizard	Vulpecula	Vulpeculae	Vul	Little Fox

NAMED STARS IN Go To CATALOGS							
Name	Designation	Name	Designation	Name	Designation	Name	Designation
Acamar	θ Eridani	Antares	α Scorpii	Hassaleh	ι Aurigae	Rasalas	μ Leonis
Achernar	α Eridani	Arcturus	α Boötis	Homam	ζ Pegasi	Rasalhague	α Ophiuchi
Acrux	α Crucis	Arkab	β Sagittarii	Izar	ϵ Boötis	Regulus	α Leonis
Acubens	α Cancri	Arneb	α Leporis	Kaus Australis	ϵ Sagittarii	Rastaban	β Draconis
Adara, Adhara	ϵ Canis Majoris	Ascella	ζ Sagittarii	Kaus Borealis	λ Sagittarii	Rigel	β Orionis
Adhafera	ζ Leonis	Asellus Australis	δ Cancri	Kaus Media	δ Sagittarii	Rigil Kentaurus	α Centauri
Albireo	β Cygni	Asellus Borealis	γ Cancri	Kocab, Kochab	β Ursae Minoris	Ruchbah	δ Cassiopeiae
Alcaid, Alkaid	η Ursae Majoris	Aspidiske	ι Carinae	Kornephoros	β Herculis	Rukbat	α Sagittarii
Alchiba	α Corvi	Atik	ζ Persei	Lesath	ν Scorpii	Sabik	η Ophiuchi
Alcor	80 Ursae Majoris	Atria	α Trianguli Australis	Markab	α Pegasi	Sadachbia	γ Aquarii
Alcyone	η Tauri	Avior	ϵ Carinae	Matar	η Pegasi	Sadalbari	μ Pegasi
Aldebaran	α Tauri	Baten Kaitos	ζ Ceti	Mebsuta	ϵ Geminorum	Sadalmelik	α Aquarii
Alderamin	α Cephei	Bellatrix	γ Orionis	Megrez	δ Ursae Majoris	Sadalsuud	β Aquarii
Alfirk, Alphirk	β Cephei	Betelgeuse	α Orionis	Mekbuda	ζ Geminorum	Sadr	γ Cygni
Algedi	α Capricorni	Biham	θ Pegasi	Menkalinan	β Aurigae	Saiph	κ Orionis
Algenib	γ Pegasi	Canopus	α Carinae	Menkar	α Ceti	Scheat	β Pegasi
Algieba	γ Leonis	Capella	α Aurigae	Menkent	θ Centauri	Sheliak	β Lyrae
Algol	β Persei	Caph	β Cassiopeiae	Menkib	ξ Persei	Shaula	λ Scorpii
Algorab	δ Corvi	Castor	α Geminorum	Merak	β Ursae Majoris	Shedar, Shedir	α Cassiopeiae
Alhena	γ Geminorum	Cebalrai	β Ophiuchi	Merope	23 Tauri	Sirius	α Canis Majoris
Alioth	ϵ Ursae Majoris	Cor Caroli	α Canum Venaticorum	Miaplacidus	β Carinae	Skat	δ Aquarii
Alkes	α Crateris	Dabih	β Capricorni	Mimosa	β Crucis	Spica	α Virginis
Almaak, Almach	γ Andromedae	Deneb	α Cygni	Mintaka	δ Orionis	Suhail	λ Velorum
Alnair	α Gruis	Deneb Algedi	δ Capricorni	Mira	o Ceti	Sulafat, Sulaphat	γ Lyrae
Alnath, Elnath	β Tauri	Deneb Kaitos	β Ceti	Mirach	β Andromedae	Talitha	ι Ursae Majoris
Alnasl, Alnazl	γ Sagittarii	Denebola	β Leonis	Mirfak, Mirphak	α Persei	Tania Australis	μ Ursae Majoris
Alnilam	ϵ Orionis	Diphda	β Ceti	Mirzam, Murzim	β Canis Majoris	Tania Borealis	λ Ursae Majoris
Alnitak	ζ Orionis	Dschubba	δ Scorpii	Mizar	ζ Ursae Majoris	Tarazed	γ Aquilae
Alphard	α Hydrae	Dubhe	α Ursae Majoris	Muscida	o Ursae Majoris	Tejat Posterior	μ Geminorum
Alphecca, Alphekka	α Coronae Borealis	Edasich	ι Draconis	Nair al Saif	ι Orionis	Thuban	α Draconis
Alpheratz	α Andromedae	Enif	ϵ Pegasi	Naos	ζ Puppis	Turais	ι Carinae
Alrai, Errai	γ Cephei	Errai	γ Cephei	Nihal	β Leporis	Unukalhai	α Serpentis
Alrescha	α Piscium	Etamin	γ Draconis	Nunki	σ Sagittarii	Vega	α Lyrae
Alshain	β Aquilae	Fomalhaut	α Piscis Austrini	Phad, Phecda	γ Ursae Majoris	Vindemiatrix	ϵ Virginis
Altair	α Aquilae	Furud, Phurud	ζ Canis Majoris	Pherkad	γ Ursae Minoris	Wazat	δ Geminorum
Altais	δ Draconis	Gacrux	γ Crucis	Polaris	α Ursae Minoris	Wezen	δ Canis Majoris
Alterf	λ Leonis	Gomeisa	β Canis Minoris	Pollux	β Geminorum	Yed Posterior	ϵ Ophiuchi
Aludra	η Canis Majoris	Graffias	β Scorpii	Porrima	γ Virginis	Yed Prior	δ Ophiuchi
Alula Australis	ξ Ursae Majoris	Grumium	ξ Draconis	Procyon	α Canis Minoris	Zaniah	η Virginis
Alula Borealis	ν Ursae Majoris	Hadar	β Centauri	Propus	η Geminorum	Zavijava	β Virginis
Ankaa	α Phoenicis	Hamal	α Arietis	Rasalgethi	α Herculis	Zosma	δ Leonis

names and the Bayer letters, so many stars have all three designations. The star Denebola, for example, in the tail of Leo, is also Beta Leonis and 94 Leonis. In general, names are used for the brighter stars only, though the lists of stars in Go To catalogs habitually use these rather than the Bayer letters, rather inconveniently in some cases. At least with the Bayer letter, one has an idea of where Delta Leonis might be, though not everyone may remember or realize that Zosma is the same object. Flamsteed letters apply only where there is neither a name nor a Bayer letter, and mostly cover stars of magnitudes 4 to 6.

Fainter stars still are referred to by their numbers in one of several catalogs, such as the Henry Draper Catalog, prefaced HD. Several commercial companies have regarded star catalogs as a source of stars which they then name after anyone who cares to pay them a fee, but these names are not recognized by any authority and there is nothing to prevent several companies selling the same star to different people. If you bought this atlas hoping to locate "your own" star, then the bad news is that the name is not genuinely recognized and the star may well be too faint to be shown on these maps. If the company has supplied you with a map, you may be able to locate its general area using this atlas, but spotting the actual star may require a fairly large telescope and some hard work linking the star's position with objects that you can find.

THE GREEK ALPHABET					
α	alpha	ι	iota	ρ	rho
β	beta	κ	kappa	σ	sigma
γ	gamma	λ	lambda	τ	tau
δ	delta	μ	mu	υ	upsilon
ϵ	epsilon	ν	nu	φ	phi
ζ	zeta	ξ	xi	χ	chi
η	eta	ο	omicron	ψ	psi
θ	theta	π	pi	ω	omega

The planets

There are four planets that can cause confusion when learning the constellations – Venus, Mars, Jupiter, and Saturn. Of these, Venus is not really a problem because it is so bright that it stands out very plainly. Jupiter is also bright, but could easily be mistaken for a bright star. Mars and Saturn are usually about the same brightness as many stars, so they can definitely create puzzlement. Mercury, the other bright planet, is only ever seen low in the twilight sky so it is usually out of the picture.

The planets always remain close to the ecliptic, and you can find their approximate positions using tables (provided on www.stargazing.org.uk). You can also pick out a planet by the fact that it rarely twinkles. This is because it has a definite disk, rather than being just a point of light as in the case of a star. It is the effect of our own unsteady atmosphere on the delicate wavefront of a star's light that causes twinkling, whereas the wavefront of light from a planet is less easily deviated.

Over a period of time, the planets move through the sky. They share in its daily movement, but have their own movements in addition. Each cycle of appearance of a particular planet is referred to as its *apparition*. Mercury and Venus are always to be found within a certain distance of the Sun as they are closer in to the Sun than is the Earth. They start each apparition in the western twilight sky after sunset, then move away from the Sun so that they become more easily visible. After what is called their *greatest elongation* from the Sun, they move back toward it, passing between the Earth and the Sun in what is known as *conjunction* – technically, the moment when a planet shares the same longitude on the ecliptic as the Sun, though it may be above or below the Sun in the sky, and in the case of Mercury and Venus either in front of or behind the Sun. After a week or two they can be seen again in the early morning sky just before sunrise, and they then repeat their movement away from the Sun and back again.

The other planets begin each apparition in the early morning sky just before sunrise. Over the months they move westward away from it, rising earlier and earlier until they are eventually visible at midnight on the meridian. At this point they are opposite the Sun in the sky, and more or less at their closest to Earth and therefore at their brightest. This point is called

▲ *In March 2004, Jupiter was in Leo (top), altering the normal appearance of the constellation (bottom).*

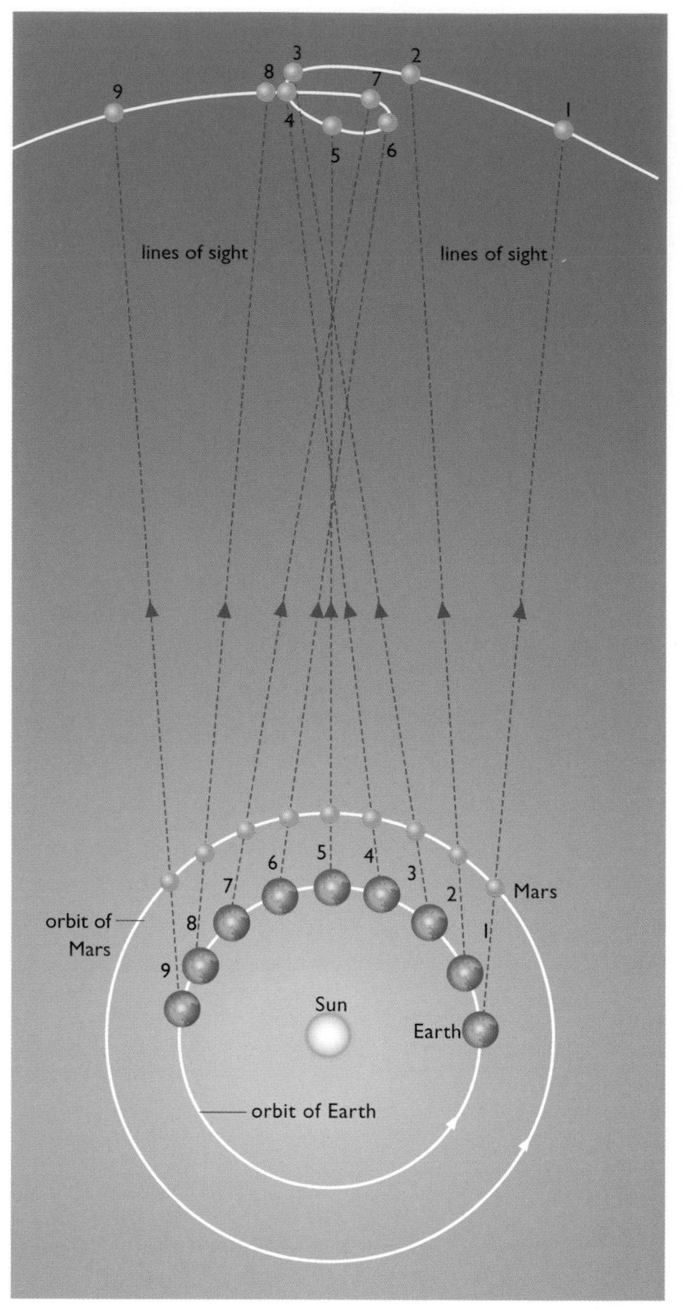

▲ *The Earth's faster motion in its orbit causes the movement of the superior planets (those with orbits that lie outside that of the Earth) to appear to reverse in a retrograde loop around the date of opposition.*

opposition. From that point they move farther into the evening sky, when most people find it easiest to observe them. However, the planets are now moving away from Earth and getting smaller in the sky, so the nearer to opposition you observe them the larger they will appear in your telescope. As time goes by they start to sink into the western sky at sunset. Finally, they pass on the far side of the Sun, at what is also termed conjunction.

These movements are the same as those of the background of stars against which the planets appear, and they are the result of the Earth's annual movement around the Sun. But each planet has its own separate eastward movement along the ecliptic, so that in the case of Mars in particular, which is moving round its own orbit but more slowly than Earth, it can take a long time for Earth to catch up with it and then pull away from it; this means that Mars' apparitions are longer than those of any other planet.

Around the time of opposition, the planets farther out from the Earth stop their eastward movement and describe slow loops in the sky, known as retrograde loops. These are the result of the Earth's own faster motion around the Sun.

Top tips for learning the sky

☆ Begin by learning the major stars and constellations using the seasonal maps and the star charts.
☆ Fill in the gaps when you know the main celestial signposts.
☆ Watch throughout the year to see all the constellations and get to know what appears when.
☆ Identify any planets using the tables found on www.stargazing.org.uk. They are the ones that do not twinkle.
☆ Look for objects of interest within each constellation using the individual charts on pages 84 to 125.
☆ Compare the appearance of stars with the naked eye and binoculars or a telescope, and get used to the additional magnification and brightness of objects when seen with optical aid.

The Milky Way

One other feature of the night sky is shown on the maps – the Milky Way. This is a pale band of light that was commonly seen by our ancestors but which is now all but invisible to most people because of light pollution. But on those occasions when you are far from civilization and have a really clear, dark sky, it appears so bright that you wonder how you could ever miss it.

The Milky Way is actually our own Galaxy of stars seen from the inside. The term may refer to both the band of light and the Galaxy as a whole. The faint, distant, individual stars merge one into another as seen with the naked eye, though binoculars and telescopes help you to pick them out. Along with the stars are dust and gas, which hide the more distant stars from view. Many of the interesting objects in the sky, such as open star clusters and nebulae, lie along the plane of the Milky Way.

Distances in astronomy

The distances to most astronomical objects are so much greater than those in everyday use that a different style of distance measurement has been adopted. Kilometers and miles can be used for distances within the Solar System, but beyond that it is commonplace to speak in terms of *light years*. A light year is the distance that light travels in one year, and is equivalent to 9.46 million million km. The distances to stars within the Milky Way range from a few light years to thousands of light years, while the distances of other galaxies are measured in millions of light years.

◄ *Fisheye view of the Milky Way in northern-hemisphere summer. It is shown from Cassiopeia in the north to Sagittarius in the south. The dark band along its central plane is caused by dust clouds and is known as the Great Rift or Cygnus Rift.*

CHAPTER 2
EQUIPMENT FOR OBSERVING

Observing with the naked eye alone goes only so far. Sooner or later, everyone wants to be able to see the heavens in greater detail, to study features on the Moon and planets, and to observe deep sky objects – the clusters and nebulae – that are barely visible, if at all, with the naked eye. This chapter is a very brief introduction to the different ways in which you can observe and record the heavens.

The naked eye

There are some characteristics of the eye that every observer should know about. The eyeball is in effect an optical instrument with an *aperture* – the hole through which light enters – a few millimeters in diameter. In bright sunlight the pupil of the eye closes down to about 1.5 mm diameter, while in the dark it opens up to between 5 and 9 mm diameter (the younger you are, the wider it can open). The detector that picks up the light is in effect a 120 megapixel receptor array, compared with the 12 to 18 megapixel arrays of the better digital cameras. Most of these receptors are called rods, and they are not color-sensitive. The rods simply detect light within a wavelength range of 380 nm to 760 nm (nm = nanometer, which is a billionth of a meter) – that is, violet to deep red. Color vision comes from different receptors called cones, which are concentrated toward the center of vision.

The receptors are more densely packed toward the center of vision, so we have only a small area of good vision. We are not normally aware of this, as our eye continually darts from place to place in order to fill in the details, but as you look at these words try to perceive details of your surroundings without looking directly at them. You will notice that you cannot see any great detail, and also that colors are not very obvious. Away from your center of vision the more sensitive rods are more common than cones, which means that night vision is better, being primarily provided by the rods. Your center of vision is the worst part of your eye for detecting faint objects, which is an obvious drawback for astronomers. For this reason, faint objects are best seen by using *averted vision* – that is, by using your peripheral vision without staring directly at the object.

The diameter of the pupil is not the only way, or even the primary way, in which the eye regulates its sensitivity. At low light levels the sensitivity of the rods is increased by a chemical called rhodopsin. This takes some time to have an effect, with the result that if you walk straight out of a brightly lit room into the dark you have virtually no night vision. Full sensitivity is only achieved after 30 to 45 minutes or maybe much longer. Rhodopsin breaks down in bright light, but is more strongly affected by blue rather than red light. For this reason, you should use a red light when looking at charts or making notes while observing.

Before you go out to observe, avoid using lighting with a strong blue or white content. TV screens, computer monitors, and fluorescent lights are bad for your night vision, but incandescent lighting has a lower blue content, particularly if you use 40-watt bulbs (though not the low-energy variety, which do have a high blue content).

There is a blind spot in each eye, located about 17° away from your center of vision on the side away from your nose. When viewing with both eyes your brain fills this spot in, but when observing through an eyepiece you will not see stars at this point.

Choosing and using binoculars

Virtually every stargazer has binoculars, even those with an array of powerful and expensive telescopes. Binoculars give a low-magnification, comparatively wide-field view of a small part of the sky. They are useful as a quick means of locating objects that are either too faint to be seen with the naked eye or are hidden by light pollution. Some of the best views of the larger star clusters and brighter nebulae are given by binoculars, and bright comets are a glorious sight. In a dark sky, the multitudes of stars in the Milky Way are breathtaking. Although binoculars are not powerful enough to show details on the planets, you can see some of the major features on the Moon. There are also many variable stars whose brightness changes can be monitored using only binoculars.

The specification of binoculars for astronomy is not as critical as might be supposed. Binoculars are described as, for example, 10 × 50 or 7 × 42, where the first figure in each case is the magnification and the second is the diameter of the main or objective lenses in millimeters. So 10 × 50 binoculars (pronounced 10 by 50) have a magnification of 10 times (10×) and objective lenses 50 mm across. A magnification of between 7 and 10 is ideal for astronomy, with 12 as an option in light-polluted skies where extra magnification helps to show fainter stars. In general, the lower the magnification, the wider the field of view and the brighter the image.

There is, however, a limit to the brightness that you can use. The figure derived from dividing the magnification into the objective diameter gives the size of the circle of light emerging from the eyepieces, known as the *exit pupil*. For 10 × 50 binoculars this is 5 mm, while for 7 × 50s it is just over 7 mm. As the pupils of older people may not open wider than about 5 mm anyway, the extra light provided by 7 × 50 binoculars can be wasted. This is also the reason why it is a waste of time making, say, 3 × 50 binoculars. No one has eyes wide enough to use all the light coming from them.

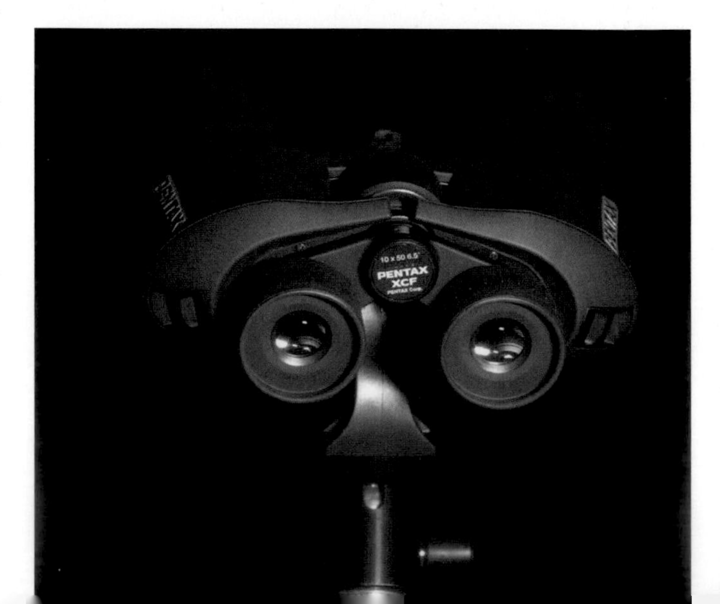

▶ *These 10 × 50 binoculars are a good compromise between magnification and convenience. A tripod helps to keep the image steady, though image-stabilized binoculars are available at a considerable price.*

The *actual* field of view – that is, the amount of sky – of specific binoculars can be between about 4° and 8°. This is not to be confused with the *apparent* field of view, which is the apparent size of the circle of light that you see when looking through and is typically between 40° and 60°. The actual field of view depends not only on the magnification (in that higher magnification generally results in a smaller field of view) but also on the nature and quality of the eyepiece in the binoculars. It is better to have a smaller, good-quality field of view than a wide one with poor definition, in which stars are not sharp. The field of view is often not specified, but may be given in the form of feet at 1000 yards. A field of view of 5° shows 261 feet of the landscape at a distance of 1000 yards.

The higher the magnification, the more difficult it is to hold the binoculars steady when viewing the sky. The magnification has no effect on the size or weight of the binoculars, which depends mostly on the size of the objective lenses. Binoculars with 50 mm lenses are heavier than 30 mm binoculars of the same type, but will show fainter stars. So the choice of binoculars depends on many factors, and it is best to test several, ideally on the night sky, before you buy. Binoculars for astronomy need not be expensive – you do not need the same level of waterproofing or robustness that a birdwatcher or mountaineer might demand – but very cheap instruments may have poor optical quality or be easily jolted out of alignment. Avoid zoom or very high-power binoculars sold cheaply, particularly through mail-order ads, even those that claim to be ideal for astronomy.

Binocular adjustments

In most binoculars, the separation of the eyepieces can be adjusted, there is a center focus wheel, and the focus of one eyepiece can be adjusted to suit individual eyes. If you wear spectacles for long or short sight, you can probably remove them when viewing because the focusing range should be adequate to allow for your vision. If you use spectacles for astigmatism or extreme focusing defects, however, you may need to use binoculars with rubber eyecups that fold flat to allow you to press them right against your spectacles. Models that allow a larger eye relief – that is, the distance between the eyepiece and your eye – may be useful in this case.

It is usually easier to adjust binoculars for your eyesight by day than by night. Begin by getting the separation of the two halves exactly right for the distance between your eyes, so you are looking directly down the center of each eyepiece. Then, using your left eye only, focus on a distant object using the center focus wheel. Finally, use your right eye only and use the adjustable eyepiece (sometimes called the diopter focuser) to focus on the same object. Most binoculars have scales for the separation of the eyepieces and the diopter correction, so make a note of these and you can quickly adjust any binoculars to suit your eyes.

Telescope choice

There are so many different types of telescope available that the choice for the beginner is almost overwhelming. As well as the three main optical systems of refractor, reflector, and catadioptric (explained below), there are also electronically controlled mounts that claim to be able to find virtually any object in the sky after just a simple setup procedure. Portability also affects people's choice, with many observers wanting telescopes that they can take to dark-sky sites or abroad.

The three main types of telescope are: refractors, which use lenses to focus the light; reflectors, which use mirrors; and catadioptrics, which use a combination of the two in a more compact design. The choice of mounting is as important as the optical design of the telescope, and in many cases you can get any type of telescope on any type of mounting, which adds to the range of possibilities from which to choose.

The basics of optics apply equally to all types of telescope. Each has a main mirror or lens whose job is to collect and focus the light from the object being observed. Main lenses tend to be called *objectives* or objective glasses, abbreviated OG. The aperture or diameter of the objective or mirror is its most important specification – the larger the diameter, the more light it collects. When we refer to a 150 mm or 6-inch telescope, we mean its diameter rather than its length.

The other important figure is the *focal length*. This is the distance between the lens or mirror and the focus point, and in ordinary refractors or reflectors it dictates the length of the telescope tube. It can be made short or long, and the different focal lengths have different properties. More important than the actual dimensions of the telescope is the comparison of the focal length with the diameter of the lens or mirror. Known as the *focal ratio* or *f-number*, it is the focal length divided by the diameter. So, in simple terms, a tube five times as long as it is wide has a focal ratio of 5, written f/5.

While a low f-number provides a compact instrument, there are penalties. One is that it is hard to achieve the same optical performance with short focal ratios compared with longer ones, because the optics require steeper curvature. Another is that it is harder to achieve high magnifications, because much of the magnification comes from the actual focal length of the

1 Adjust the separation between the eyepieces to equal the distance between your eyes.

2 Focus on a distant object using the left eyepiece and the center wheel.

3 Adjust the focus of the right eyepiece using the diopter focuser.

main lens or mirror. There is a practical limit of about f/4 for reflectors and f/5 for refractors. In general terms, telescopes with short f-numbers are better suited to deep sky observing in dark skies, while those with long f-numbers are more suited to planetary observing or observing in towns.

Eyepieces and magnification

Every telescope needs an eyepiece to make the image observable and to provide magnification. The eyepiece itself has a focal length, and the overall magnification of a telescope is found by dividing the focal length of the telescope by that of the eyepiece. So a telescope of 500 mm focal length, used with an eyepiece of 10 mm focal length, will give a magnification (or power, as it is often called) of 50. To achieve a magnification of 100, you need an eyepiece of 5 mm focal length. The same eyepieces used with a telescope of 1000 mm focal length would give magnifications of 100 and 200. In general, eyepieces are available in the range 4 mm to 40 mm.

Barlow lenses are devices that sit between your telescope and the eyepiece; they multiply the effective focal length of the telescope, usually by a factor of 2. If you are building a set of eyepieces, try to get a range that does not include factors of 2 in focal length, so that a 2× Barlow will increase the number of steps in your range. For example, if you already have a 26 mm eyepiece, a 2× Barlow will give you a 13 mm equivalent. There is little point in buying a 12 mm, as this is too close, so instead buy a 15 mm to give you a range of four widely separated focal lengths with just three purchases.

Increased magnification has two drawbacks: the field of view becomes smaller, and the image of an extended object becomes dimmer, as the same amount of light is spread over a larger area. The maximum magnification that you can use is limited by two factors – the steadiness of the atmosphere and the properties of light itself. Larger telescopes are often limited by the first of these factors to powers of about 200. Small instruments are limited by the wave nature of light, which means that only so much detail is visible through a particular aperture. A rule of thumb is that the maximum usable magnification is about twice the aperture in millimeters. A 60 mm refractor is therefore limited to a magnification of about 120.

At one time there were several different types of eyepiece available, each with its own advantages and drawbacks. Today, the Plössl type is ubiquitous as a basic eyepiece, except on the cheapest refractors, which use the simpler Huygenian eyepieces. There are also specialist eyepieces available, notably those that provide very wide fields of view. These are often very bulky and expensive, and can easily weigh and cost more than a starter telescope.

▲ *A range of Plössl eyepieces in $1\frac{1}{4}$-inch fittings, from 4 mm (bottom left) to 40 mm (top right).*

Eyepieces generally have a barrel diameter of 31.7 mm, more usually described as $1\frac{1}{4}$ inches. Some small refractors may use 24.5 mm barrels. Extra-wide-field, long-focal-length eyepieces may require a barrel of 50 mm (2 inches) in order to give the full field of view, but these can only be used on telescopes with the right size focusing mount.

It is a basic fact of optics that virtually all astronomical telescopes give an inverted view – that is, upside down. In the past, it was always said that astronomers would rather have an upside-down image than put extra bits of glass in the light path in order to bring the image the right way up. These days, however, many people spend large sums of money on eyepieces with extra glass elements that give a wide field of view. But having a non-inverted or erect image is not deemed a priority, other than with smaller telescopes that are also intended for daytime use.

Refractors

These are what most people think of as a telescope, with the main lens at the top of the tube and the eyepiece at the bottom. The smallest telescopes are all refractors, and they are available with apertures up to about 150 mm at reasonable cost. At one time, refractors were typically about f/12 or longer, but today it is not unusual to find f/5 instruments of good quality.

▼ *A 102 mm f/4.9 refractor on AZ3 altazimuth mount with manual slow motions. Notice the star diagonal, which gives easier viewing angles. This instrument has a "star pointer" zero-power finder rather than a finder telescope.*

Refractors have always suffered from the defect known as *chromatic aberration* or false color, which means that not all colors of light are focused at the same point. The effect when viewing a planet is of a bluish halo surrounding the planet's disk, though otherwise the image in a good refractor has high contrast and brightness. Filters can reduce the false color, but only very expensive refractors using what are called *apochromatic* lenses are free from it. It is only these apochromatic refractors that are really suitable for photography, and they also have particularly wide fields of view of good definition.

A refractor will not deteriorate significantly over time, and any dust that accumulates on the lens can be easily removed with care.

When observing an object high in the sky, the viewing angle can be difficult. For this reason, refractors are usually supplied with star diagonals, which turn the image through 90° and provide a much more comfortable viewing angle. However, these diagonals give a mirror image of the object, and this must be taken into account when comparing observations with those from other instruments.

Reflectors

The standard reflector type is the Newtonian, in which the eyepiece is located at the top of the tube and at right angles to it. Though this means observing at right angles to the direction of the object, the result is often a fairly comfortable viewing angle without the need for star diagonals. With long-focal-length instruments, however, the eyepiece can be inconveniently high from the ground.

The mirrors are coated on the front surface, rather than on the back as with household mirrors, so the surface coating is delicate and prone to deterioration. If the tube is capped when not in use, the surfaces should last for years. Reflecting systems are more sensitive to correct alignment than refractors, so adjustments are provided, and the telescope may need realignment or *collimation* from time to time. However, mirrors give perfectly color-free images so they are equally suited to both visual and photographic observing. For details of collimation refer to www.stargazing.org.uk.

If you want the largest aperture for a set budget, a Newtonian reflector is the way to do it. In terms of actual observing versatility, a reflector offers the greatest chance of observing the largest number of objects, whether you live in a town or in the country. A moderately large aperture will still show more objects than a smaller one even in the town, but it will perform better in darker skies.

As the tube of a reflector is usually open at the top end, tube currents caused by local temperature variations can swirl round inside and reduce the steadiness of the image. Reflectors with open tubes avoid this, but they are susceptible to extraneous light, and the mirrors are prone to dewing and collecting dust. A reflector will not perform well until it has reached the same temperature as its surroundings. If you take it straight from a warm room into the cold night air, the image will be very disturbed until it has cooled down, and it may need an hour or so at outside temperature before it performs well. Refractors and catadioptrics do suffer from the same effect, but to a lesser extent.

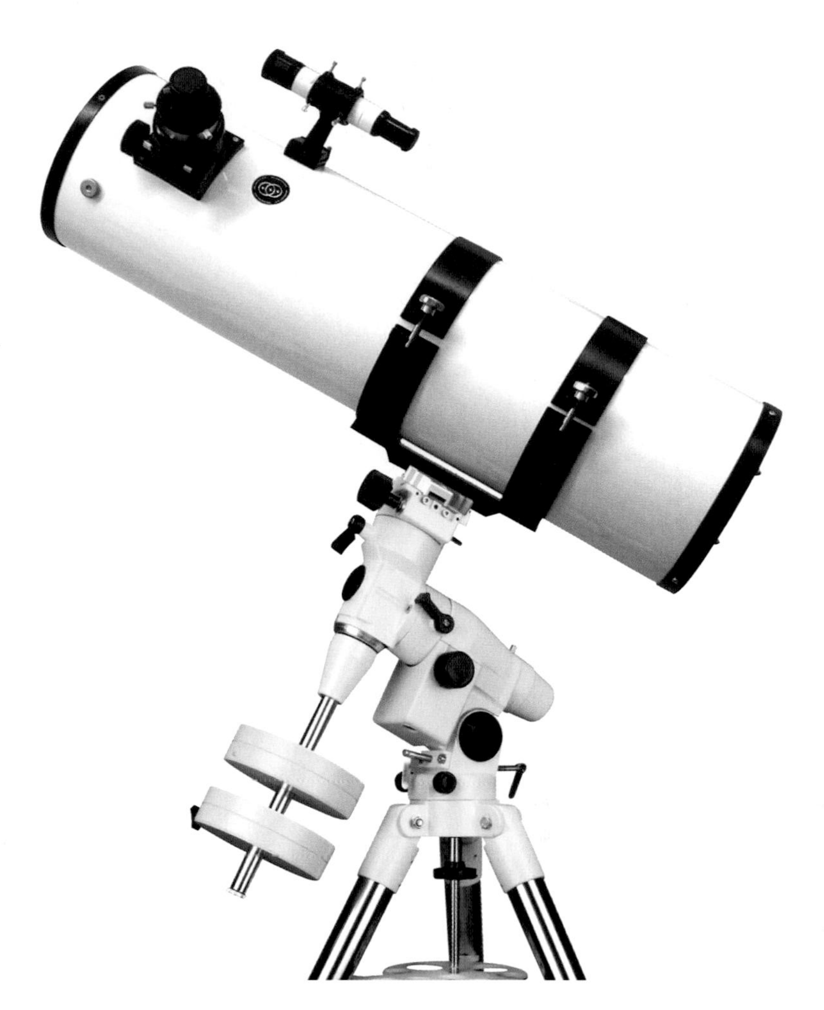

▲ *A 200 mm f/4.5 Newtonian reflector on a German-type equatorial mount. The tube can be rotated within the cradle to bring the eyepiece to a convenient viewing angle. It has a 6 × 30 finder telescope.*

Catadioptric telescopes

These are the most popular instruments available nowadays for serious amateur astronomy, eclipsing the more traditional refractors and reflectors. They include the types known as Schmidt-Cassegrain and Maksutov-Cassegrain, the latter often simply called a Maksutov. Catadioptric telescopes are basically short-focus reflectors with lenses at the top end to correct for the defects of the image. So they combine short tubes with good optical quality.

The fact that their image quality is not quite as good as that from a refractor or Newtonian reflector is outweighed for most people by their much greater compactness. This means that a 200 mm instrument is the norm, and 250 mm and even 300 mm instruments can be transported fairly easily, whereas a Newtonian of that size would require a large tube and heavy mounting. Smaller instruments are usable on a tabletop and can be carried with one hand, unlike their more traditional counterparts.

Schmidt-Cassegrain telescopes, often abbreviated to SCTs, are usually f/10, while Maksutovs are usually between f/13 and f/15. Maksutovs are particularly suited to planetary observing. They are more compact than SCTs, but require more careful manufacture and are therefore more expensive.

Another design, the Schmidt-Newtonian, is sometimes classed as a catadioptric, but this uses a small lens near the eyepiece rather than a large one at the top of the tube. Its

▲ *A basic 200 mm f/10 Schmidt-Cassegrain telescope on a fork mount. This has the same aperture as the reflector shown on the previous page but has a much shorter tube. The telescope comes equipped with motors and Go To controller.*

performance is generally slightly poorer than the conventional Newtonian of the same aperture. For more details on telescope choice, refer to www.stargazing.org.uk.

Finders

Finding an object in the sky using even a small telescope magnifying, say, 30 times, is amazingly difficult compared with finding an object on land. All you see is blank sky, with no landscape features you might recognize. The solution is to use a finder, and even most Go To telescopes need a finder to help you set them up at the beginning of a session.

There are two basic types: small refracting telescopes mounted parallel to the main instrument but giving a low magnification and a wide field of view, and red-dot finders, which appear to project a red dot on the sky when you look through them, but with no magnification. These are now commonly found on budget telescopes as they are cheap to make and tend to be more usable than the very poor finder scopes that were previously common.

Aligning the finder perfectly with the main telescope is a vitally important first step, but one that many beginners ignore. It is best done during daylight by finding a distant object with the main telescope, then adjusting the finder so that the same object is precisely on its crosswires or dot. Without this step, finding objects at night can be frustrating at best and doomed to failure at worst.

To find any object in the sky, first look along the telescope tube to get the right part of the sky. This is often harder than it seems, and manufacturers are surprisingly lax in providing simple sighting devices on telescopes to help the beginner. With practice, you should be able to point your telescope at a chosen object, which should then be well within the field of view of the finder. Then bring the object to the center of the finder's crosswires or dot and if it is properly aligned, the same object will be in the low-power field of view of the main telescope.

Telescope mounts

The mounting of a telescope is often crucial to its performance. Many cheap telescopes are let down by their flimsy mountings, and would be more usable if they were remounted. There are two basic types of mount – the altazimuth and the equatorial.

Altazimuth mounts

The name of these mounts refers to their two movements, one in altitude (up and down) and one in azimuth (side to side). At one time these mounts were restricted to only the very cheapest of telescopes, but today they are found on virtually all sizes, from the very smallest through to the largest professional telescopes, the advent of computer control overcoming their drawbacks.

They are usually of the fork or yoke type, in which the telescope pivots up and down between the arms or tines of a twin fork, which itself rotates about a central axis. This is simple and effective in engineering terms, but the drawback is that celestial objects move through the sky at an angle (as described in Chapter 1). This produces two problems: first, it means that the telescope has to be moved on both axes simultaneously; and second, the plane of the image rotates as the object moves through the sky. Though not a great problem for visual observers or even for short-exposure imaging, it rules out long-exposure photography.

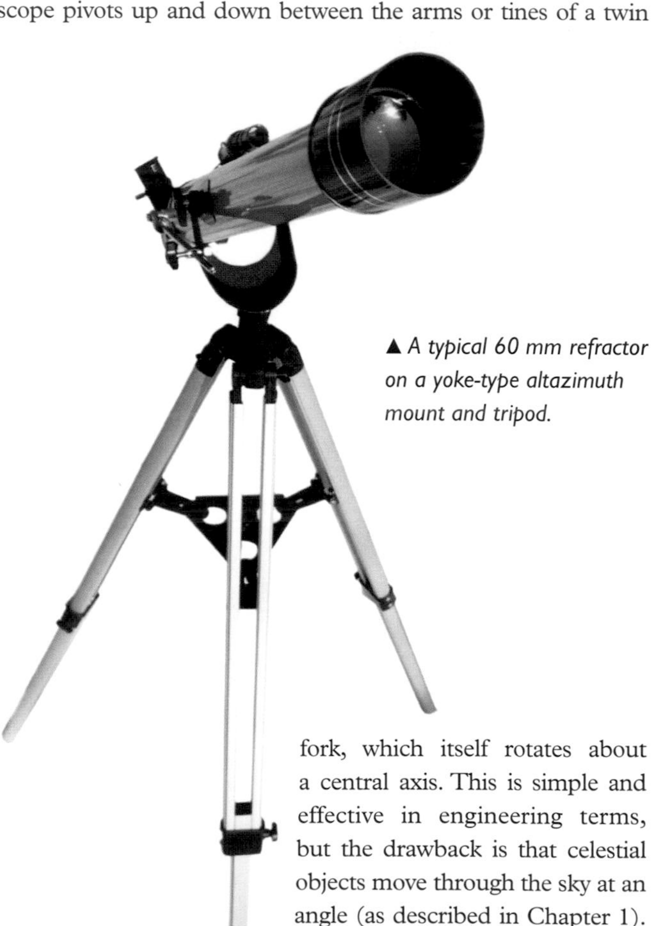

▲ *A typical 60 mm refractor on a yoke-type altazimuth mount and tripod.*

Modern altazimuth mounts overcome the first problem by using a purpose-built computer handset. It contains a virtual model of the whole sky, and once it knows its orientation compared with just two fixed points in the sky it can find any other

▼ *An example of field rotation. As the constellation Taurus moves through the sky its orientation changes. The inset shows a close-up of the Pleiades star cluster at the given times. Even if you tracked the center of the cluster perfectly, a time exposure would still have star trails around the center.*

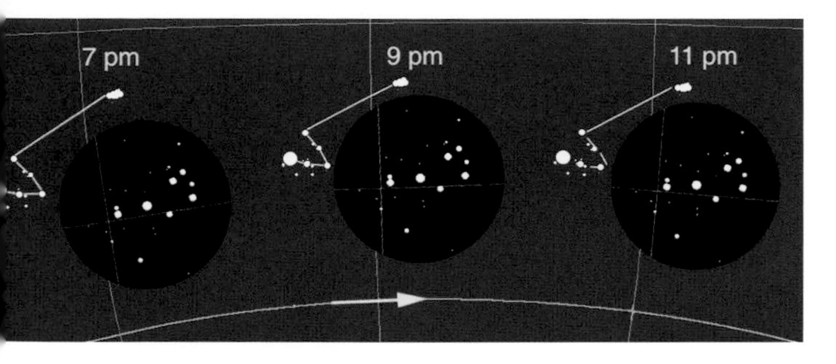

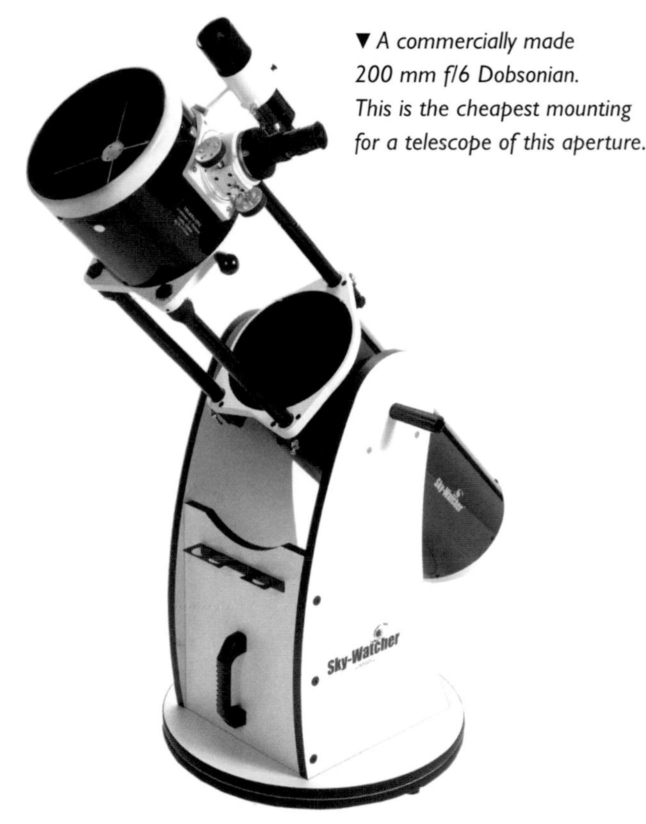

▼ *A commercially made 200 mm f/6 Dobsonian. This is the cheapest mounting for a telescope of this aperture.*

◀ *The Meade ETX Maksutov telescopes use altazimuth fork mounts with a computer-control option to provide a Go To facility.*

object whose position it knows. This is the Go To facility. A Go To computer, however, does not overcome the second problem – that of image rotation. It is possible to buy image derotators, which turn the camera during the exposure, but these are not commonly used.

The engineering simplicity of the altazimuth mount is exploited in the Dobsonian mount. The traditional Dobsonian is made from wood, with Teflon pads between the rotating surfaces to reduce friction. It is intended as a cheap but effective mounting and is particularly suited to the larger sizes of Newtonian telescope. Dobsonians offer the largest aperture for your money, and they are favored by deep sky observers who are interested in using big telescopes to find faint objects. They can also be used for casual planetary observing, but are not ideal as they cannot conveniently be motorized and cannot be used for long-exposure photography.

Equatorial mounts

The drawbacks of the altazimuth mount are overcome by the equatorial mount. It again has two axes of motion, but one of them is inclined parallel to the Earth's axis. In the case of a fork mount this simply involves tilting the base at an angle equal to the latitude of the observing site using what is usually referred to as a wedge. The telescope can then be driven using one motor only to follow any astronomical object's motion through the sky. The main drawbacks of the fork mount are that it is not suited to long tubes, and that the eyepiece becomes inaccessible when it moves between the arms of the fork.

The most common alternative to the fork mount, the German mount, overcomes both these deficiencies but has its own foibles. One is that it requires a counterweight to the telescope, which adds to the weight of the whole assembly, and the other is that full rotation across the sky is not usually possible without some part of the assembly hitting some other part, or the whole telescope needing to be turned to the other side of the mount and rotated through 180°.

To set up an equatorial mount for ordinary observing, rather than for photography, simply set the polar axis at the correct angle for your latitude – there is usually a latitude scale marked on the mount or on the wedge. Make sure that the base is level, then point the polar axis due north in the northern hemisphere or south in the southern. This approximate alignment should be adequate for most purposes, and anything you observe should remain within the field of view for a considerable time with only minor corrections for drift. For long-exposure photography, however, you will need much more stringent and advanced polar alignment (beyond the scope of this introductory text).

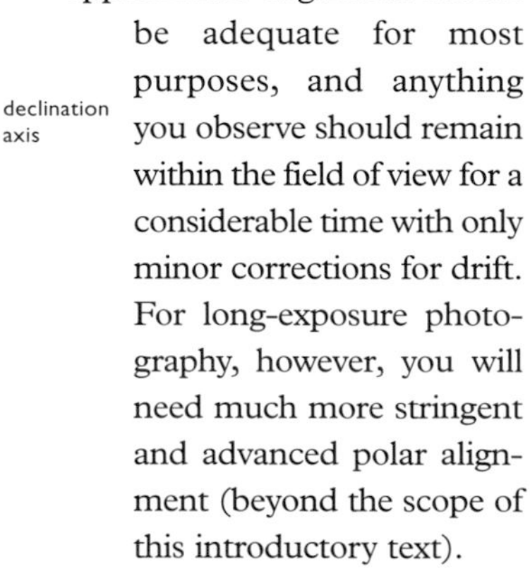

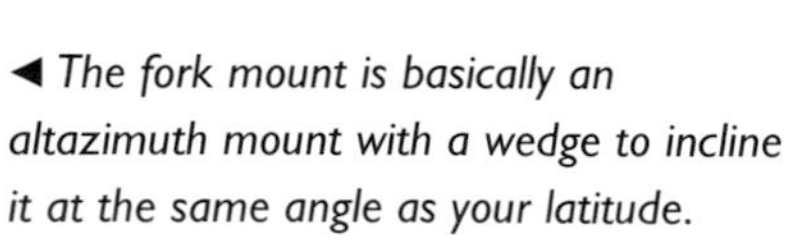

◀ *The fork mount is basically an altazimuth mount with a wedge to incline it at the same angle as your latitude.*

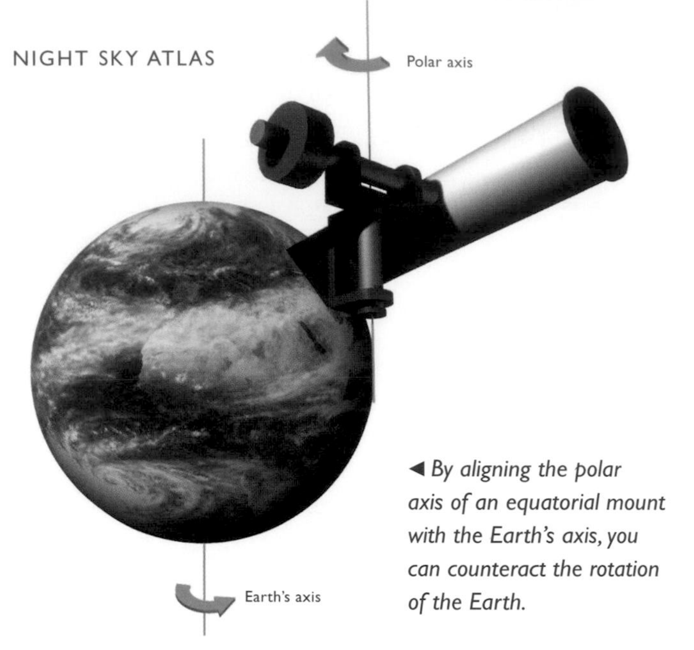

◀ *By aligning the polar axis of an equatorial mount with the Earth's axis, you can counteract the rotation of the Earth.*

Equatorial mounts are usually provided with setting circles, which are graduated in right ascension and declination. The declination scale is fixed, but the RA scale can be rotated to suit the time of observation. The easiest way to use the scales is simply to point the telescope at a star whose position you know. If the mount is correctly polar aligned the declination scale should be correct, and you can simply turn the RA scale to read the correct position. Then you can find another object whose position you know using the circles alone. Some circles run in the opposite direction to this, in which case you need to work out the difference in RA between the known object and the one you want to observe, and turn the telescope accordingly.

Motor drives

Useful for all types of observing, motor drives are now widespread and are virtually essential for imaging. Once you have found an object it stays in the field of view. But even the most perfectly engineered and set-up mounting is likely to introduce small errors into the drive rate, so for long-exposure imaging some means of monitoring and correcting the drive rate is needed, a process known as guiding. At one time this was done by eye alone, watching a star and making manual corrections to keep it on crosswires. This meant using either an off-axis guider, which directed a small part at the edge of the incoming light beam to a separate eyepiece, or a guide telescope attached to the main instrument. Today, most people use separate guide telescopes, but instead of the tedium of continually watching a star, they use autoguiders which incorporate an electronic imager to monitor errors and make corrections.

Imaging

The word "photography" is now often replaced in the astronomical vocabulary by "imaging." The end result is still a photograph, but "imaging" tends to imply a longer process than simply pressing a shutter button. Though it is possible to take a picture through a telescope simply by holding the camera up to the eyepiece and taking a snap, the results are usually rather hit-or-miss. This is known as the *afocal method*, and it is the only method possible with compact cameras or digital cameras with non-removable lenses. With special clamps to hold the camera in place, you can get very successful photos, but there are severe limitations with this method. The camera is usually quite heavy, which may result in balance problems; operating the shutter often introduces vibration unless you can use a remote control lead or cable release; and monitoring the drive rate is not usually possible.

In addition, most compact cameras are not designed for long exposures and are often restricted to an exposure time of up to about a minute. This is not a problem for the Moon and planets but is less useful for deep sky objects.

A revolution in planetary photography occurred when around 2000 it was realized that webcams – simple video cameras that are intended for use with a computer – provide a very simple and effective camera for imaging bright objects through a telescope. They are cheap and lightweight, and produce a video stream that can be stored for later processing. One problem that has frustrated observers for centuries is the unsteadiness of the Earth's atmosphere – usually referred to as the *seeing*. This causes a planetary image to shimmer and wobble under all but the rarest of steady conditions, but the eye and brain work together to pick out the detail. A single exposure on a camera simply records one distorted image, but the beauty of a webcam is that it can record many images per second. Software automatically picks out the best frames and adds them, creating a final image from only the best of hundreds or even thousands of frames.

This has revolutionized planetary and lunar imaging, and amateur astronomers at sea level can now take images of bright objects which far exceed what was possible even in the 20th century using large mountaintop telescopes. But commercial webcams themselves have now moved on and have become more difficult to adapt for telescope use than the early models. Instead, amateur astronomers use specialist cameras based on the same chips as those in webcams. Some units give color images directly, but the best results come from mono cameras which are used with separate red, green, and blue filters to produce a color image after image processing. However, webcam-type cameras are not usually ideal for the long-exposures needed for deep sky objects. And a computer is needed, with the attendant problems of providing a power supply during long sessions outdoors.

For long-exposure imaging through the telescope you usually need either an SLR (single-lens reflex) camera or a CCD camera. An SLR camera has a detachable lens, so the telescope can be used in place of the normal lens to provide the image. Traditionally these were film cameras, but today digital SLRs (DSLRs) are the norm. A readily available adapter is all that is needed to link telescope and camera.

As mentioned in Chapter 1, the view of most deep sky objects through even a large amateur telescope is not a patch on a

▲ *A lightweight Point Grey Flea 3 camera attached to a telescope. The unit is a fraction of the weight of even a compact camera, and gives a 648 × 488 pixel video stream in either mono or color.*

▲ *A single frame from a 2-minute webcam sequence of Jupiter, made using a 200 mm reflector, shows only a few details.*

▲ *Using Registax to combine the best 900 images from 1200 frames, and with image processing, a stunning amount of detail is visible.*

long-exposure photograph. Realizing this, many amateur astronomers now spend most of their time taking images rather than looking through the telescope. Everyday DSLR cameras can take impressive shots, far exceeding what was possible using film in the last century. They can go on building up light for minutes or hours, with the exposure time usually being limited by light pollution in all but the most remote country districts. But electronic noise or spurious speckles can often spoil the results.

For serious long-exposure photography, the cooled CCD camera reigns supreme. The cooling of the CCD chip reduces the noise level, and allows very long exposures. Image processing at the computer can remove the background light pollution, if any, and enhance faint details. As a result, it is possible to make images of faint galaxies even from city locations, while in dark skies images are possible that not long ago would have required a giant telescope.

It is not always necessary to use a telescope for long-exposure astro-imaging. Any camera that will take time exposures can be mounted piggy-back on a telescope with a driven equatorial mount, or even directly on the mount. The cradles for carrying telescopes often include a threaded stud that will carry a camera for this reason. You can take photos of constellations, or, using telephoto lenses, of the larger deep sky objects, in this way.

Telescope observing tips

The primary rule of telescopic observing is always to start with a low magnification. The standard eyepiece supplied with many telescopes has a focal length of about 25 or 26 mm, and this is the ideal eyepiece to use at the start of any observation. It gives a fairly wide field of view, so if the object you are observing is not centrally placed in the field, you should be able to spot it anyway. Furthermore, low magnifications give brighter images and are easier to focus than high powers. For the beginner, a low-power eyepiece is usually easier to look through because it has a larger exit pupil, whereas with high-power eyepieces it can be very tricky to get your eye in the right place.

Small Schmidt-Cassegrain telescopes in particular can be difficult to focus because their focusing knobs are small and require a lot of turning to make much difference. Also, unlike a Newtonian or refractor, you can't see the position of the focuser, so you have no idea which way to turn the knob. It's a good idea to find a really bright object to start with so that you can easily see which way the focusing is going. The aim is to make the blurred image as small as possible for perfect focus.

Once you have found the object you can proceed to using higher powers. Changing eyepieces takes time. If your telescope does not have a motor drive, notice how fast the object moves through the field of view and allow for this. Usually the focus position is different with the higher power, so you must refocus – which may mean that you lose the object you wanted to observe. Getting used to this needs practice.

Higher powers mean dimmer images, smaller fields of view and quicker movement of the object through the field of view if the telescope is not driven. One advantage of higher powers, apart from the obvious one of increased magnification, is that the sky background is dimmed more than are star images, which should remain as points of light. So if you want to locate a small, faint cluster, for example, using a higher magnification should improve your chances in a light-polluted area. But a faint galaxy or planetary nebula will be dimmed equally with the background by increased magnification.

Increasing the magnification of a planetary or lunar image may show more detail but it will also increase any effects of seeing. This local turbulence can be caused by heat within the telescope or its immediate surroundings, by nearby landscape features or by higher-level atmospheric turbulence. Observing from inside a warm house through an open window is particularly bad, as the warm air pours out of the window so that you are trying to observe through a shimmering air stream. You should always take your instrument outside and let it get down to night-time temperature before beginning to observe.

People often confuse seeing with *transparency* by referring to "good seeing" when they actually mean a very clear sky. Frequently, one excludes the other. A crystal clear night is often accompanied by turbulence, whereas haze can often mean steady conditions and good seeing. Some observers find that they get the best views of the planets when they are barely visible through haze or cloud. The all-round skygazer chooses the best objects to observe depending on the conditions – either clear nights for deep sky work or hazy ones for the planets.

How telescopes perform

You often need to know the limitations of your telescope. Very often these are set by the seeing conditions and the presence of light pollution, but tests do give practical values for the faintest star visible and the finest detail visible with various sizes of telescope. The *resolving power* of a telescope is a measure of the finest detail it will show, and is limited by its aperture. A small telescope will not show fine detail, no matter how much you increase the magnification – you just magnify a blur. The figure given for resolving power is the separation of the closest double star that it will show as two individual stars, but this serves as a reasonable guide to the size of detail that should be visible on planets. The table gives figures for a range of popular telescope apertures.

RESOLVING POWERS FOR DIFFERENT APERTURES

Aperture of telescope in mm	60	80	100	114	125	150	200	250
Faintest star visible (magnitude)	11.6	12.2	12.7	13.0	13.2	13.6	14.2	14.7
Resolving power (arc seconds)	1.9	1.4	1.2	1.0	0.9	0.8	0.6	0.5

SEASONAL SKY MAPS

How to use these maps

These maps provide a quick reference to the stars in your sky at any time. First choose the map that best matches the time and location at which you will be observing. From anywhere in the northern hemisphere, use the charts on this and the next page. From the southern hemisphere, use the charts on pages 24–25. The exact view depends on your latitude, so notice the lines at the bottom of each map, which indicate the position of the horizon for various latitudes. The farther south you are, the more you see of the southern part of the sky and the less you see of the northern.

The maps are in pairs, showing the view looking south and the view looking north on a particular date. Because of the movements of the sky, each pair of maps shows the sky at different times depending on whether you are viewing in the evening or the early morning, so choose the date and time closest to when you are observing. Bear in mind that these times do not allow for Daylight Saving Time (Summer Time), so the maps show the sky for an hour later than the times given when this is in force. So if you are observing at 10.30 pm in summer, use the map for 9.30 pm.

Each map shows the horizon along the bottom edge and the overhead point at the top. The edges of each semicircle are due east and west, which means that the semicircles join each other. Any constellation that crosses the edge will be split between the two halves of the map. Each map covers a wide area of sky, and a constellation that appears small on the map will cover quite a large area in reality.

Constellations move from west to east as the year progresses, so for months later than those shown, the stars at the eastern edge of the maps will be closer to the center.

Once you have chosen the best map for your observing date and time, note the page numbers at the edge of the chart to find out which of the main atlas maps will give you a larger-scale view of the same area. Remember that the planets move from month to month so they are not shown on the maps. A bright planet could change the pattern of the constellations, but you can check the positions of the planets for each month using the tables on www.stargazing.org.uk.

Use the descriptions below to find the "signpost" constellations in each part of the sky; these are the star patterns that will help you to find your way.

The northern-hemisphere sky over the seasons

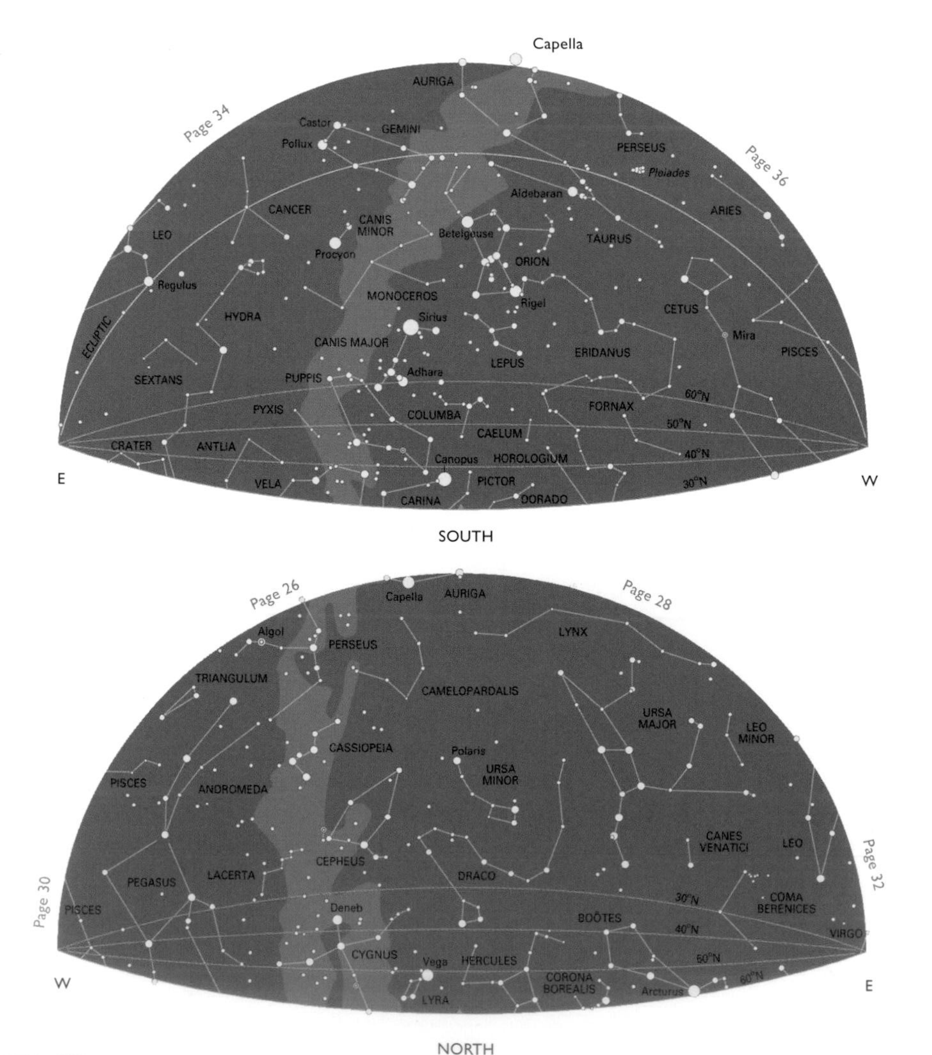

January evening signposts

Orion dominates the view at this time of year. Its bright belt of three stars is a brilliant pointer to other constellations. To the upper right they point toward the bright star Aldebaran in Taurus, and then on to the Pleiades star cluster. To the lower left they indicate Sirius, the brightest star in the sky, in Canis Major. Above Orion, roughly overhead, is the pentagon of Auriga, while to the east of this is a pair of stars known as Gemini, the Twins. Directly below Gemini lies the star Procyon in Canis Minor.

A line from Gemini through Auriga takes you to Perseus, which lies in the Milky Way. Farther on from this is the W-shape of Cassiopeia. Notice that the ecliptic is high in the sky at this time of year, so there may be bright planets in Gemini, above Orion or in Taurus.

Evening	**Morning**
January 1 at 11.30	*October 1 at 5.30*
January 15 at 10.30	*October 15 at 4.30*
January 30 at 9.30	*October 30 at 3.30*

May evening skies

On May evenings the sky looking south is particularly barren of bright stars. This is because the Milky Way lies low on the horizon and in the main part of the sky we are looking at right angles to the plane of the Galaxy. There is only one really bright star, yellowish Arcturus, in the otherwise rather faint constellation of Boötes. Below this, in mid sky, is Spica, in Virgo, with the rest of the constellation extending in a straggling Y-shape toward Regulus, the brightest star in Leo, to the west. Any planets will be in mid sky, in Virgo or Leo. The ecliptic extends toward the southeast where the reddish Antares in Scorpius is rising.

Looking north, the Big Dipper or Plough is virtually overhead, while over in the northeast the bright star Vega is a taste of the stars of summer.

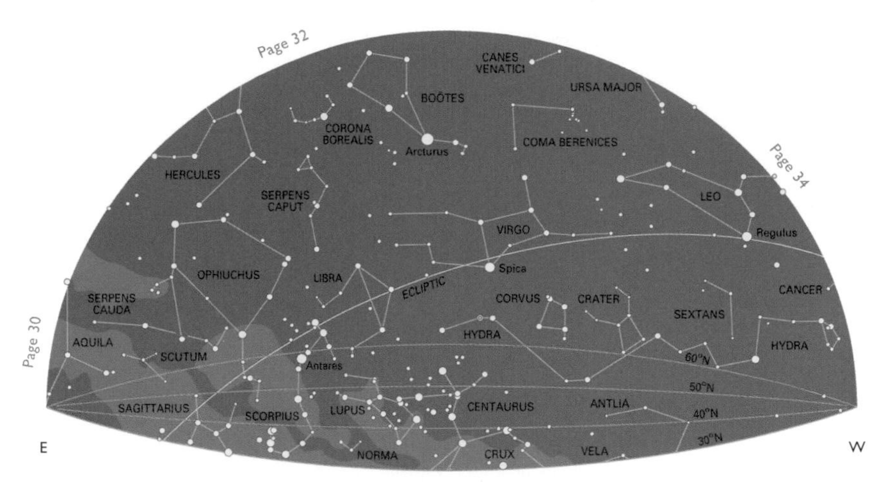

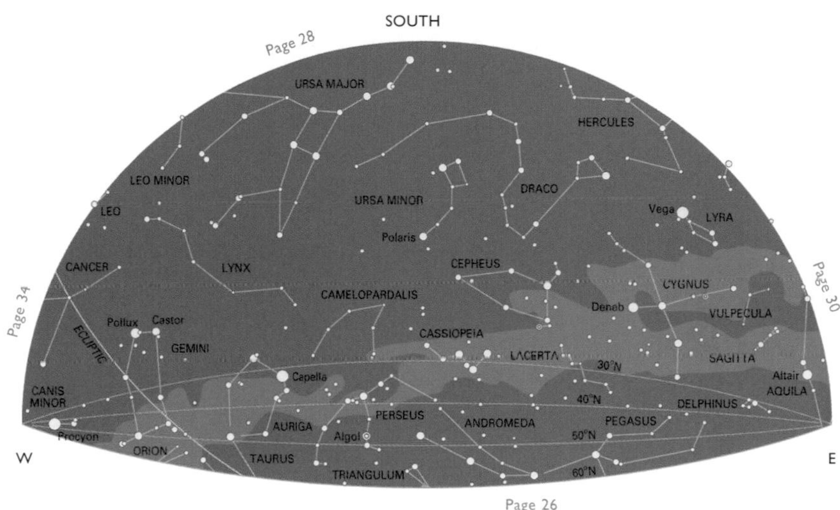

Evening	*Morning*
May 1 at 11.30	*January 15 at 6.30*
May 15 at 10.30	*February 1 at 5.30*
May 30 at 9.30	*February 14 at 4.30*

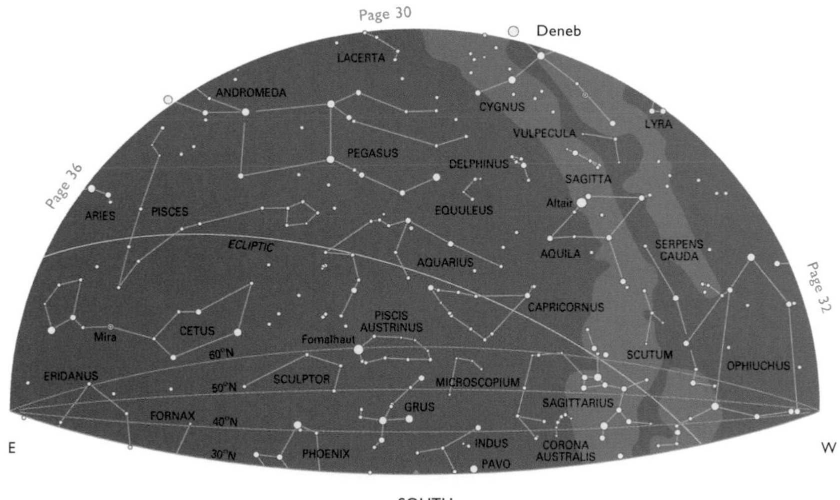

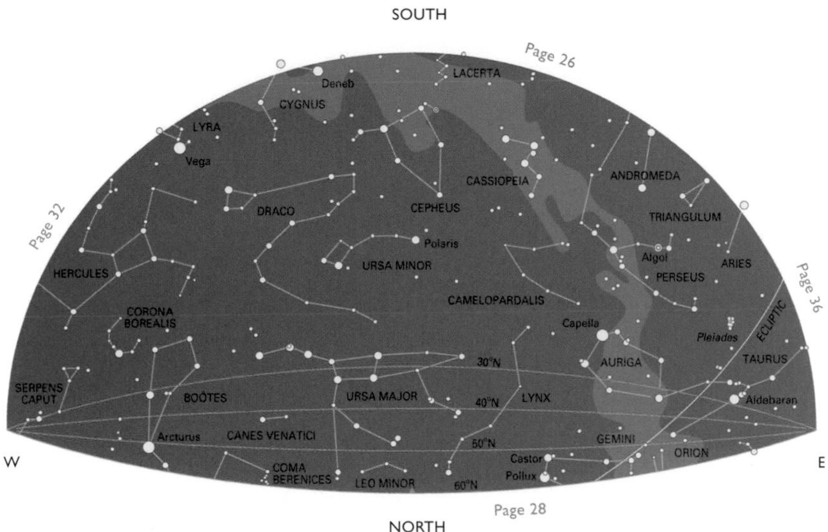

September evening skies

By September, the Milky Way is high in the sky. We are now looking more toward the center of the Galaxy, so it is much more obvious than it was in January. Deneb in Cygnus, the Swan, is at the top of a large cross-shape known as the Northern Cross. This extends along the Milky Way with Altair in Aquila, flanked by two fainter stars, to one side of the line. Follow the line southward to find Sagittarius, near the horizon, which has no very bright stars.

High in the south is the large but fairly faint Square of Pegasus, which you can use as a signpost to the faint constellation of Aquarius and the lone bright star Fomalhaut. From the upper left of Pegasus extends a widely spaced line of three stars in Andromeda. Any planets will be low in the west or in mid sky in the south. To the north, the Big Dipper or Plough is close to the horizon.

Evening	*Morning*
September 1 at 11.30	*June 15 at 4.30*
September 15 at 10.30	*June 30 at 3.30*
September 30 at 9.30	*July 15 at 2.30*

The southern-hemisphere sky over the seasons

The maps on these two pages are for use by observers in the southern hemisphere only. Full instructions for using them are given on page 22. Each pair of maps shows the sky looking either south or north on a particular date, so choose the map that is closest to the date and time when you will be observing. Individual horizons are shown for latitudes between 35°S and 5°S. The maps are also usable farther south than 35°S, such as in the South Island of New Zealand, but some stars in the lower half of the north-facing map will be below the horizon. Bear in mind that from most populated areas your actual horizon will be the upper horizon line on the north-facing map.

As before, use the page numbers around the edge of the maps to help you to choose the appropriate atlas map. However, from the southern hemisphere the view of atlas maps 3–6 is with north at the bottom, so they are seen upside down.

Although different stars are visible at any instant from different longitudes around the Earth, at a particular local time they are the same. So at 9 pm in South Africa it will be 4 am in Western Australia and different stars will be visible. But at 9 pm local time on the same date, observers in both locations will see the same part of the sky.

The southern skies are particularly noted for their large number of stars, though in terms of the brightest stars there is not much to choose between the two hemispheres. In fact, the southern hemisphere has more stars of medium brightness, particularly in the most southerly part of the Milky Way. Virtually all these stars are within our own spiral arm of the Galaxy. The Sun is within a huge doughnut of bright stars formed some 60 million years ago, and we happen to be closer to the edge that lies in the direction of Carina and Crux. The result is that the southern skies are richer in stars than those of the north.

January evening skies

Although the Milky Way stretches overhead, this is its faintest part and it is not easy to see. But the brilliant stars of Orion act as a great signpost. The three belt stars point downward and to the left to the star Aldebaran in Taurus; continuing the line takes you to the unmistakable star cluster of the Pleiades. Take the line from the belt in the opposite direction and you come to the brightest star in the sky, Sirius, in Canis Major, high overhead. Below Orion lies the pentagon of Auriga with its brightest star, Capella, at its base.

At right angles to the line of the belt, down to the right, lie the two bright stars of Gemini, the Twins. Farther on, rising in the east, are the stars of Leo. The ecliptic runs from Leo, through the faint stars of Cancer, then through Gemini and below Orion to Taurus, so the patterns in this part of the sky may be distorted by any bright planets that happen to be around.

By coincidence, the two brightest stars in the sky are in this region of sky. Sirius is practically overhead, while Canopus is farther south. Canopus is much brighter in real terms than Sirius, but it is also much more distant, at about 300 light years compared with only 8.6 for Sirius. In fact, Canopus is the brightest star in our local neighborhood of the Galaxy. Over to the southwest lies another bright star, Achernar, and roughly midway between Canopus and Achernar is the Large Magellanic Cloud, a small companion galaxy to our own. It is visible as a misty patch. The Milky Way becomes brighter toward the south, with numerous medium-brightness or faint stars defining its track.

Evening	*Morning*
January 1 at 11.30	*October 1 at 5.30*
January 15 at 10.30	*October 15 at 4.30*
January 30 at 9.30	*October 30 at 3.30*

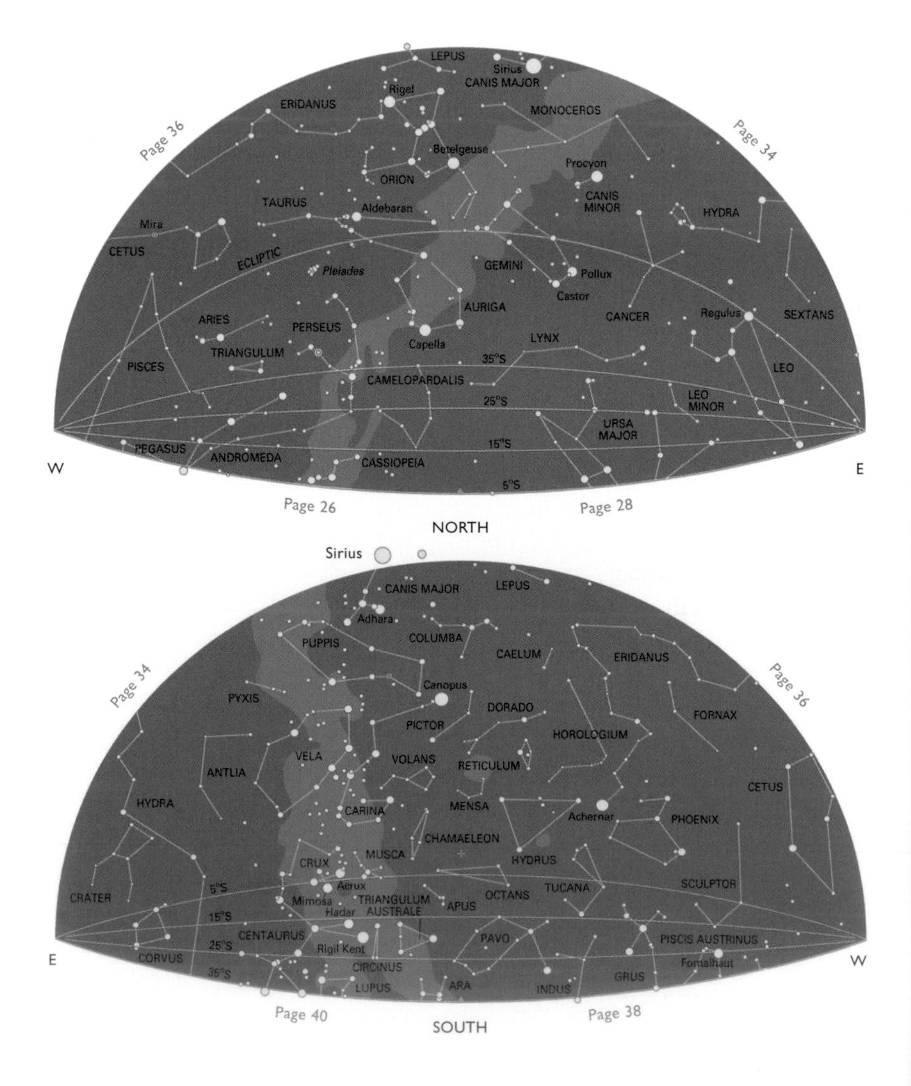

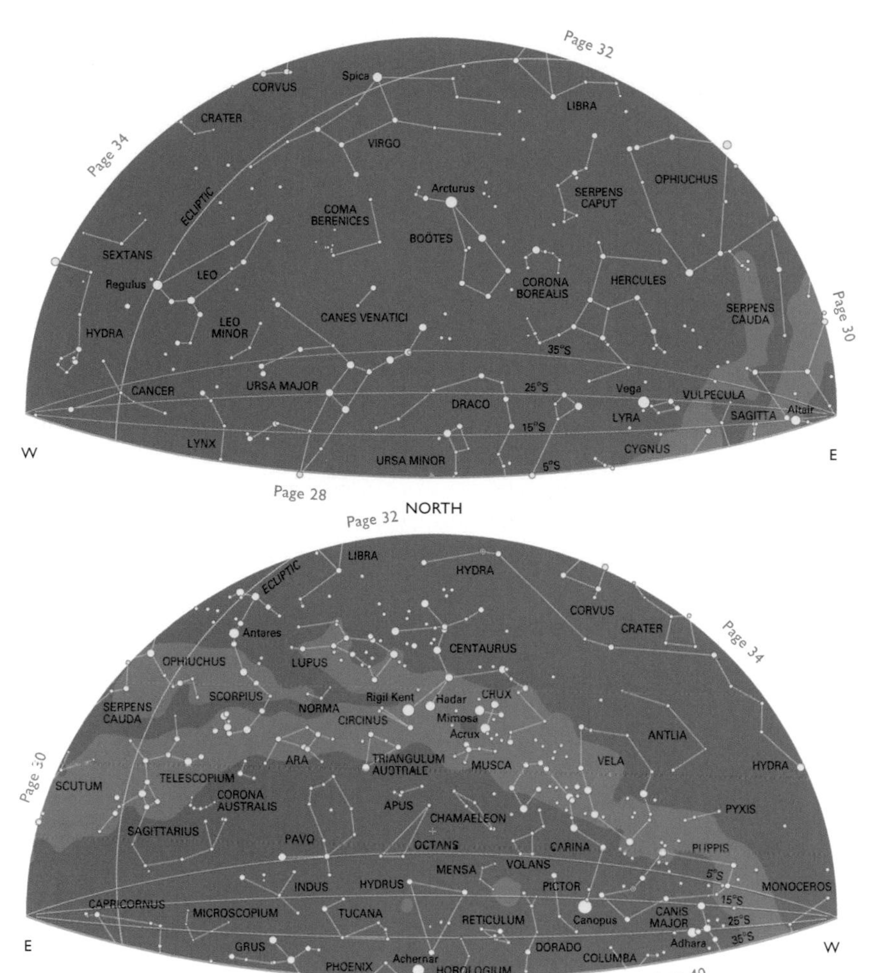

May evening skies

At this time of the year, the northern skies are devoid of bright stars other than Arcturus, in Boötes. High overhead is Spica, the brightest star in Virgo. The rest of the constellation stretches down to the west, where you will find Regulus in Leo. This also marks the line of the ecliptic, which continues east from Spica to the bright reddish star Antares, at the heart of Scorpius, the Scorpion, with the curve of its sting in the Milky Way. Beyond that lies Sagittarius, rising in the east. Any bright evening planets at this time of the year are likely to be high in the sky and well placed for observation.

If the northern skies are barren, the far southern section of the Milky Way is on full view, with large numbers of stars through Vela, Carina, Centaurus, and Crux. Many of these stars are quite hard to organize into patterns, the great exception being the Southern Cross (or Crux) and the two adjacent stars Alpha and Beta Centauri. Below and to the left of these stars is the well-defined triangle of Triangulum Australe.

Evening	***Morning***
May 1 at 11.30	*February 14 at 4.30*
May 15 at 10.30	*February 28 at 3.30*
May 30 at 9.30	*March 15 at 2.30*

September evening skies

Looking north there are few bright stars, but the Square of Pegasus, though not bright, is easy to pick out. The lower edge of the square points down to the northeast and the line of three widely spaced stars that are the brightest in Andromeda. To the east of this lies a small group of three stars that form Aries, the Ram. Take a diagonal line upward through the Square to find Aquarius, which is another constellation of the zodiac, yet with rather faint stars. Its most identifiable feature is an arrow-shape known as the Water Jar. The ecliptic continues through faint Capricornus to the more obvious stars of Sagittarius.

Low in the north is the bright star Deneb in Cygnus, also known as the Northern Cross. The Milky Way forks here in what is called the Great Rift, running past Altair to Sagittarius and Scorpius. This is the most brilliant part of the Milky Way, and the minor constellations of Vulpecula, Sagitta, and Scutum are worth finding for their deep sky riches.

Evening	***Morning***
September 1 at 11.30	*May 15 at 6.30*
September 15 at 10.30	*June 1 at 5.30*
September 30 at 9.30	*June 15 at 4.30*

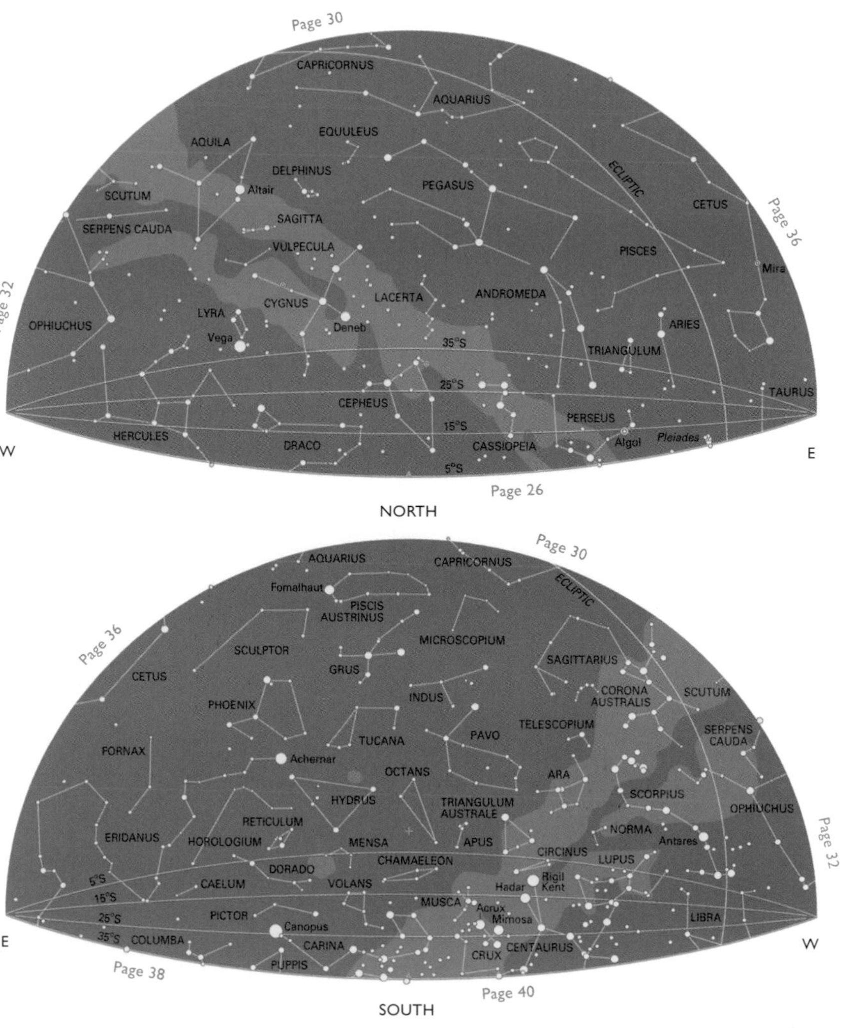

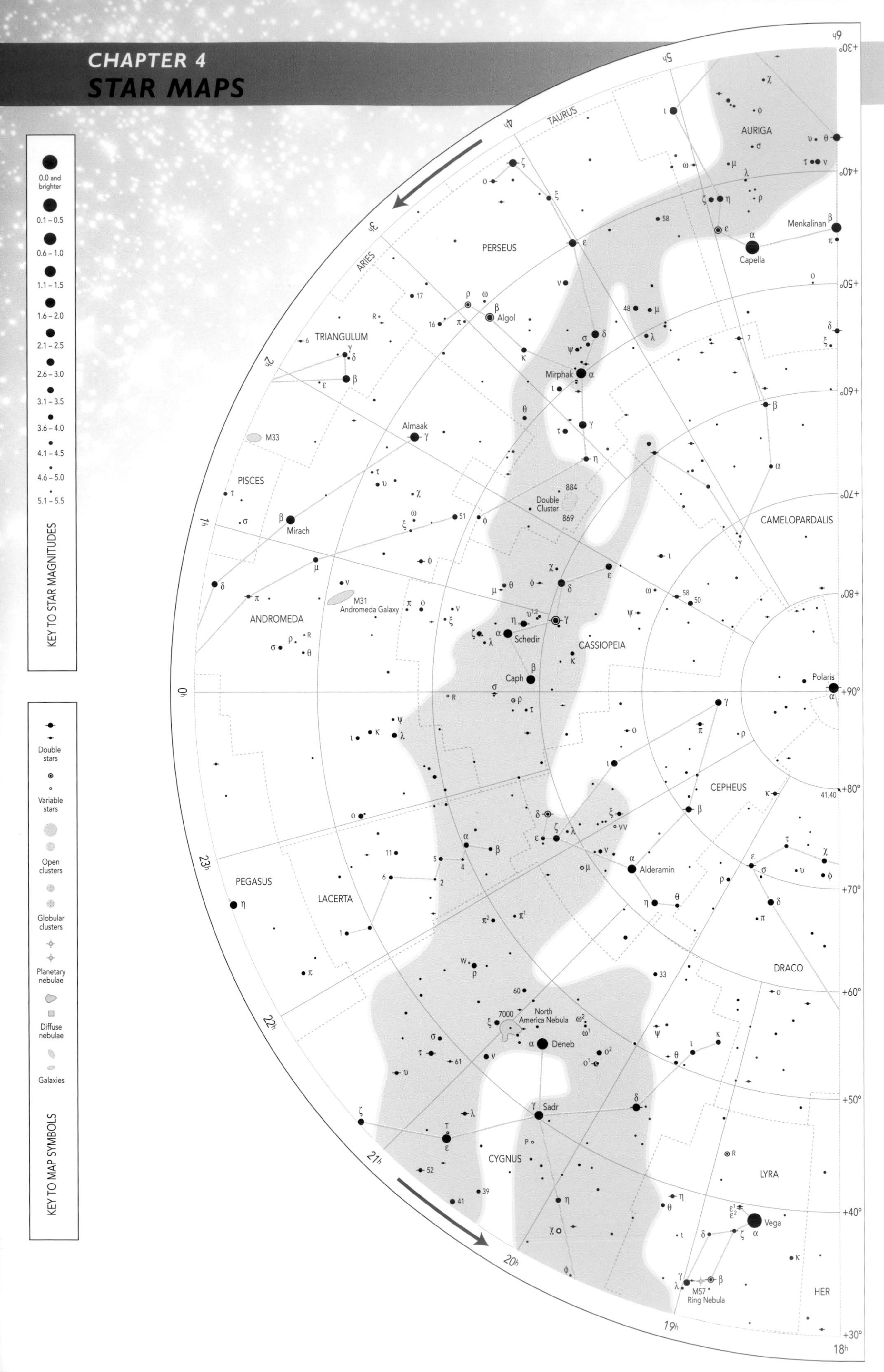
0.0 and brighter
0.1 – 0.5
0.6 – 1.0
1.1 – 1.5
1.6 – 2.0
2.1 – 2.5
2.6 – 3.0
3.1 – 3.5
3.6 – 4.0
4.1 – 4.5
4.6 – 5.0
5.1 – 5.5
KEY TO STAR MAGNITUDES
Double stars
Variable stars
Open clusters
Globular clusters
Planetary nebulae
Diffuse nebulae
Galaxies
KEY TO MAP SYMBOLS
AURIGA
Capella
Menkalinan
TAURUS
PERSEUS
Algol
Mirphak
ARIES
TRIANGULUM
M33
PISCES
Almaak
Mirach
Double Cluster
884
869
CAMELOPARDALIS
ANDROMEDA
M31
Andromeda Galaxy
CASSIOPEIA
Schedir
Caph
Polaris
CEPHEUS
Alderamin
PEGASUS
LACERTA
DRACO
North America Nebula
7000
Deneb
Sadr
CYGNUS
LYRA
Vega
M57
Ring Nebula
HER
+30°
+40°
+50°
+60°
+70°
+80°
+90°
0h
1h
2h
3h
4h
5h
6h
18h
19h
20h
21h
22h
23h

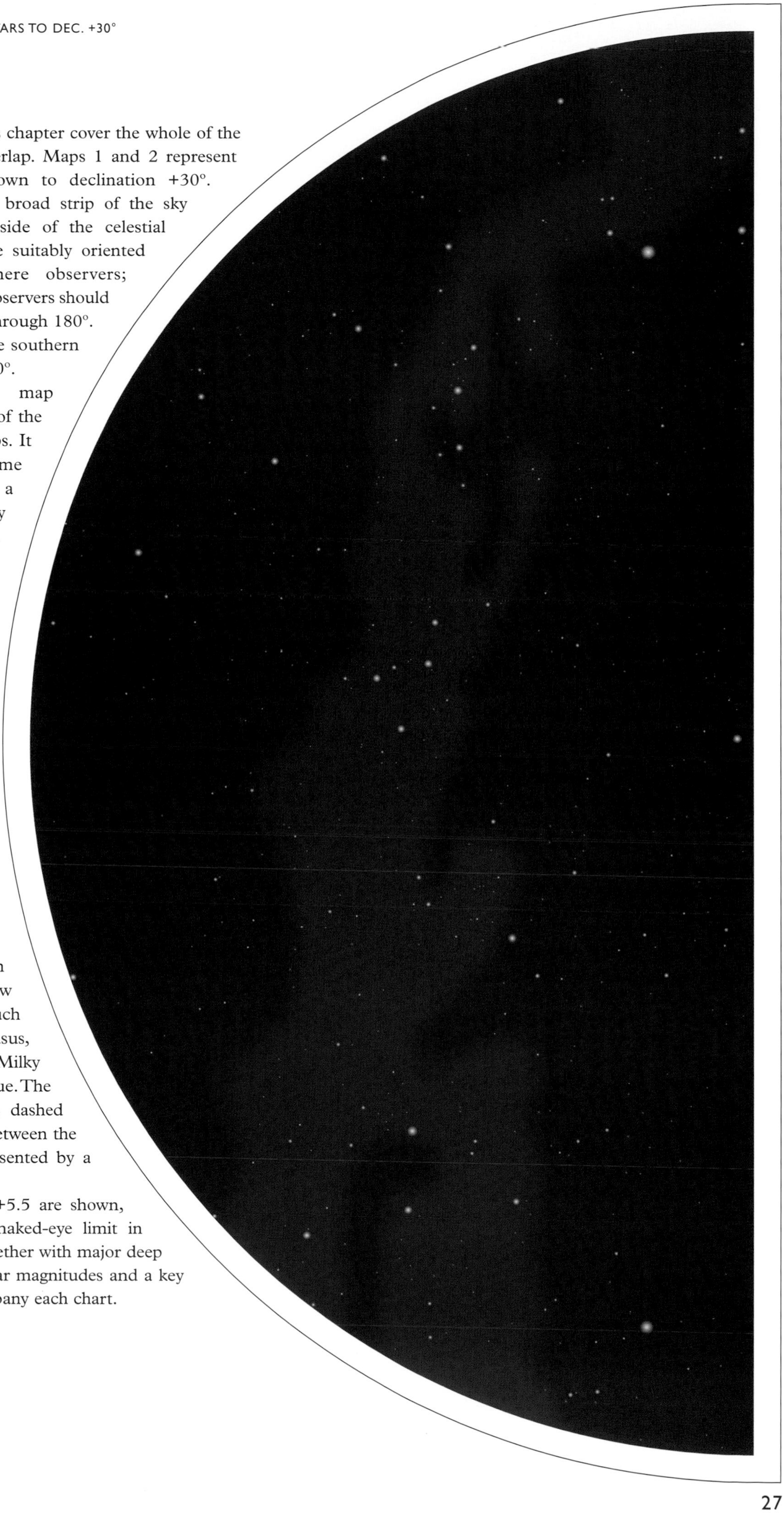

The star maps in this chapter cover the whole of the sky, with some overlap. Maps 1 and 2 represent the northern stars down to declination +30°. Maps 3 to 6 cover a broad strip of the sky extending 60° either side of the celestial equator. The maps are suitably oriented for northern-hemisphere observers; southern-hemisphere observers should simply turn the book through 180°. Maps 7 and 8 show the southern stars, to declination –30°.

A photo-realistic map appears opposite each of the more conventional maps. It shows exactly the same region of sky, but in a manner that closely resembles what you will actually see. You can match this map with your view, and then use the conventional map to find the names of the stars and constellations.

All 88 constellations are shown on these maps. An index listing on which map(s) each constellation appears can be found on page 41. The constellation names are in capital letters, while star names are in lower case with an initial capital. A few prominent asterisms, such as the Square of Pegasus, are also named. The Milky Way is shown in light blue. The ecliptic is shown as a dashed red line. The borders between the constellations are represented by a dashed pink line.

Stars to magnitude +5.5 are shown, which is roughly the naked-eye limit in semirural locations, together with major deep sky objects. A key to star magnitudes and a key to map symbols accompany each chart.

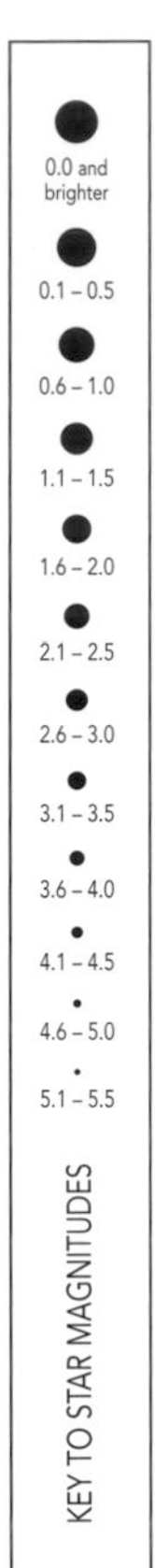
0.0 and brighter
0.1 – 0.5
0.6 – 1.0
1.1 – 1.5
1.6 – 2.0
2.1 – 2.5
2.6 – 3.0
3.1 – 3.5
3.6 – 4.0
4.1 – 4.5
4.6 – 5.0
5.1 – 5.5
KEY TO STAR MAGNITUDES

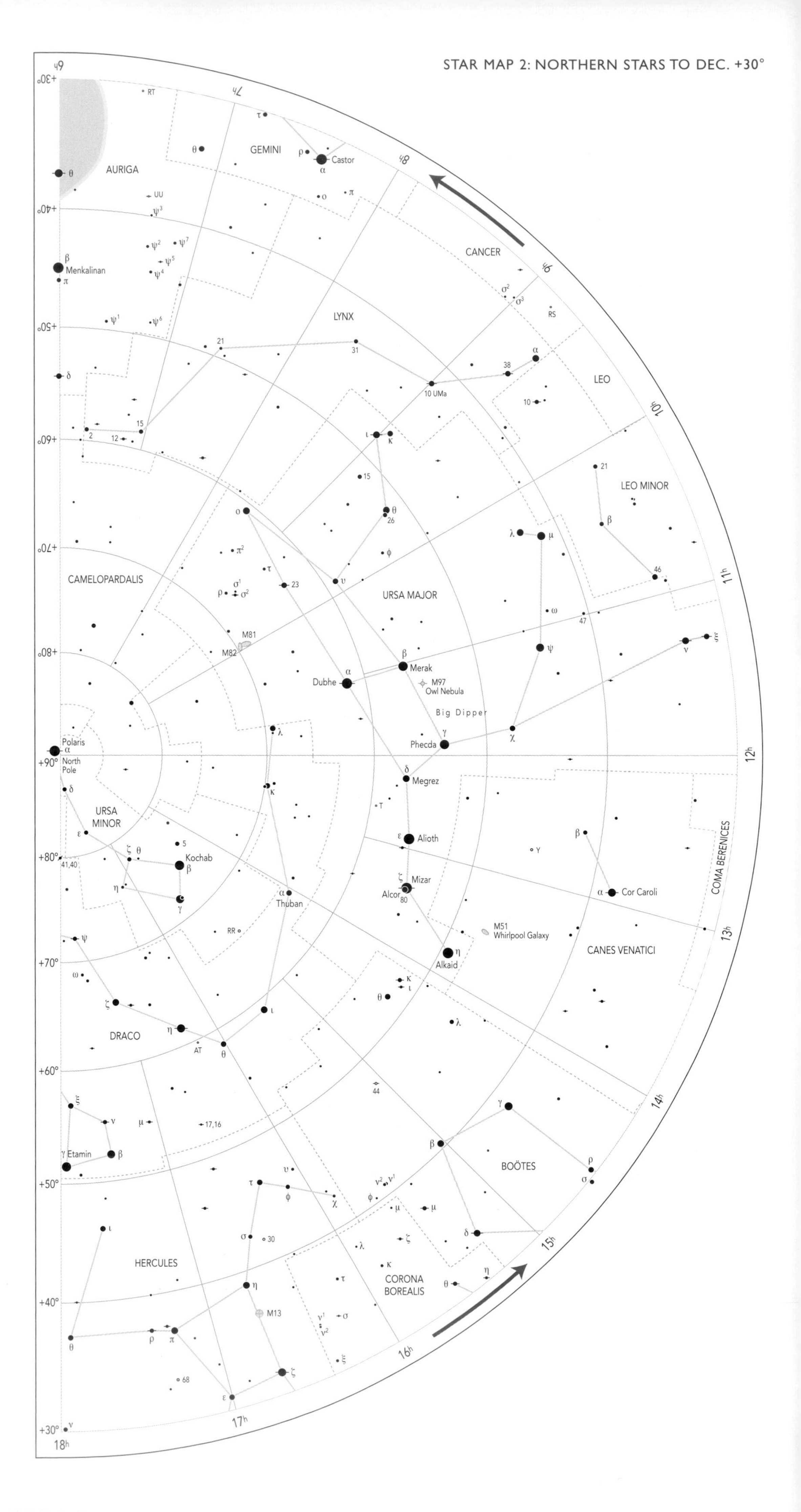
AURIGA
GEMINI
Castor
CANCER
LYNX
LEO
LEO MINOR
CAMELOPARDALIS
URSA MAJOR
M81
M82
Dubhe
Merak
M97
Owl Nebula
Big Dipper
Phecda
Megrez
Alioth
Mizar
Alcor
Alkaid
M51
Whirlpool Galaxy
CANES VENATICI
Cor Caroli
COMA BERENICES
Polaris
North Pole
URSA MINOR
Kochab
Thuban
DRACO
Etamin
HERCULES
M13
CORONA BOREALIS
BOÖTES
Menkalinan
10 UMa
6h
7h
8h
9h
10h
11h
12h
13h
14h
15h
16h
17h
18h
+30°
+40°
+50°
+60°
+70°
+80°
+90°

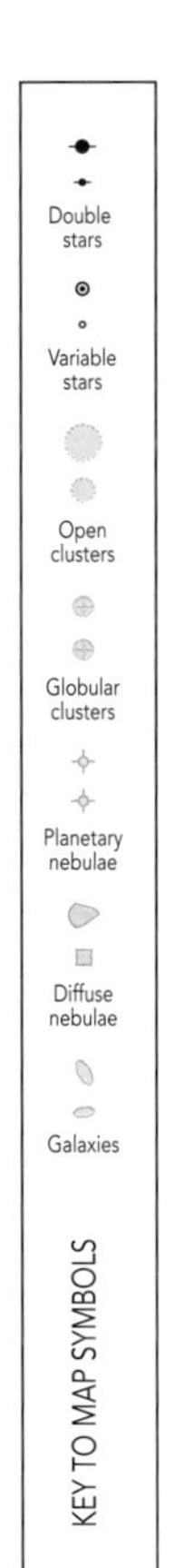
Double stars
Variable stars
Open clusters
Globular clusters
Planetary nebulae
Diffuse nebulae
Galaxies
KEY TO MAP SYMBOLS

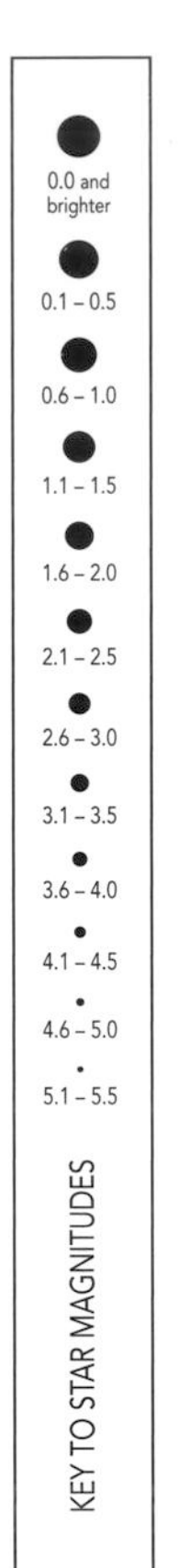
0.0 and brighter
0.1 – 0.5
0.6 – 1.0
1.1 – 1.5
1.6 – 2.0
2.1 – 2.5
2.6 – 3.0
3.1 – 3.5
3.6 – 4.0
4.1 – 4.5
4.6 – 5.0
5.1 – 5.5
KEY TO STAR MAGNITUDES

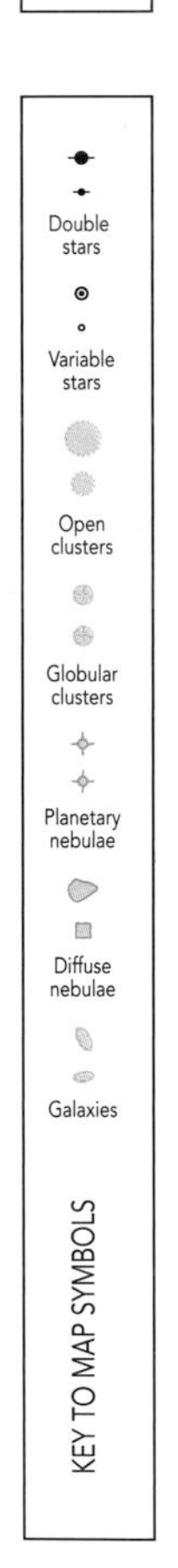
Double stars
Variable stars
Open clusters
Globular clusters
Planetary nebulae
Diffuse nebulae
Galaxies
KEY TO MAP SYMBOLS

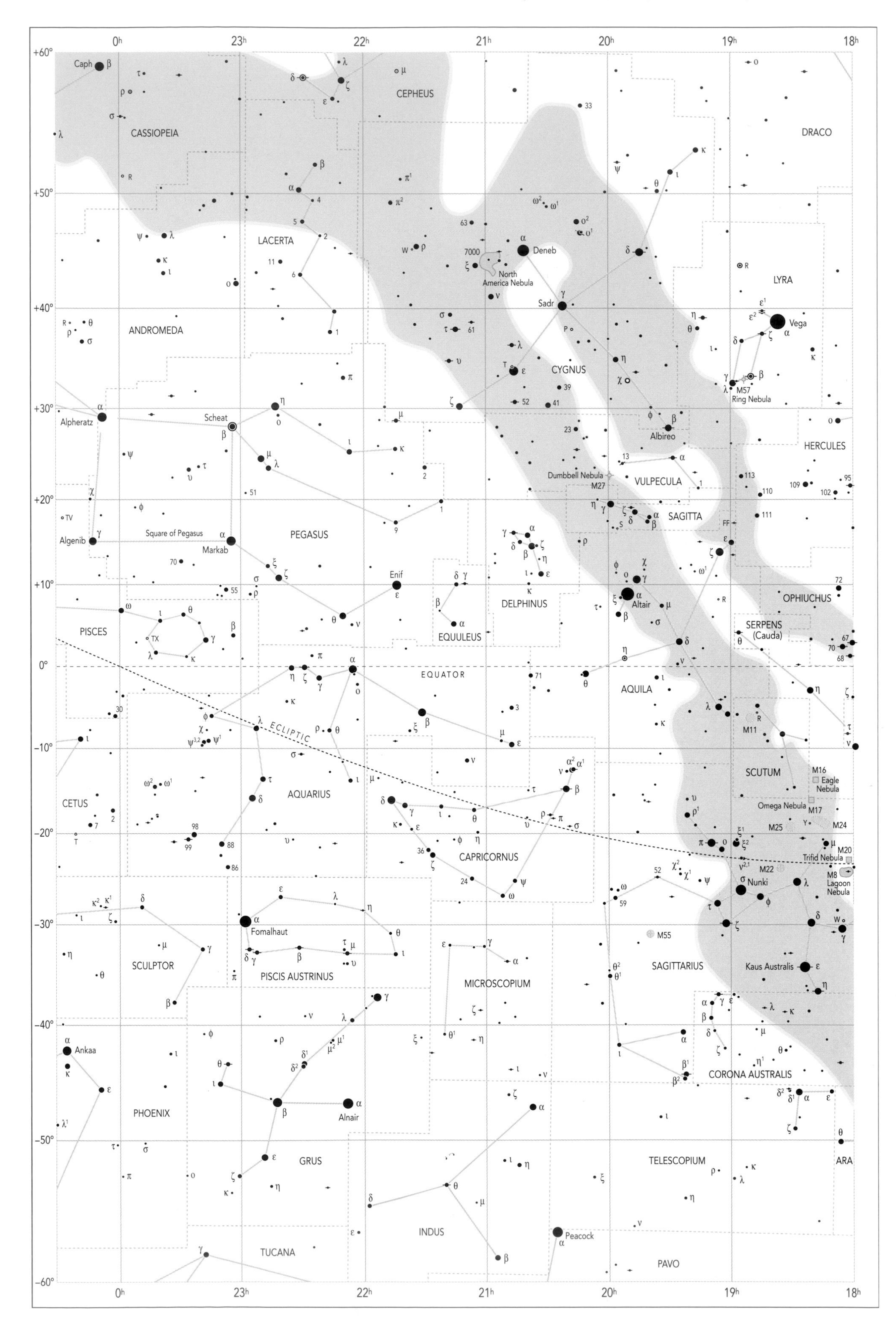
CASSIOPEIA
CEPHEUS
DRACO
LACERTA
LYRA
ANDROMEDA
CYGNUS
HERCULES
VULPECULA
SAGITTA
PEGASUS
DELPHINUS
OPHIUCHUS
PISCES
EQUULEUS
SERPENS (Cauda)
AQUILA
EQUATOR
ECLIPTIC
CETUS
AQUARIUS
SCUTUM
CAPRICORNUS
SCULPTOR
PISCIS AUSTRINUS
MICROSCOPIUM
SAGITTARIUS
PHOENIX
CORONA AUSTRALIS
GRUS
TELESCOPIUM
ARA
INDUS
TUCANA
PAVO
Caph
Deneb
North America Nebula
Sadr
Vega
M57 Ring Nebula
Alpheratz
Scheat
Albireo
Dumbbell Nebula
M27
Square of Pegasus
Algenib
Markab
Enif
Altair
M11
M16 Eagle Nebula
Omega Nebula M17
M25
M24
M20 Trifid Nebula
M22
Nunki
M8 Lagoon Nebula
Fomalhaut
M55
Kaus Australis
Ankaa
Alnair
Peacock

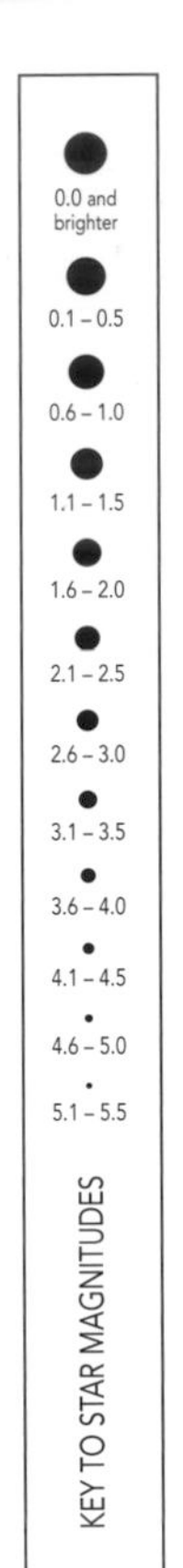
0.0 and brighter
0.1 – 0.5
0.6 – 1.0
1.1 – 1.5
1.6 – 2.0
2.1 – 2.5
2.6 – 3.0
3.1 – 3.5
3.6 – 4.0
4.1 – 4.5
4.6 – 5.0
5.1 – 5.5
KEY TO STAR MAGNITUDES

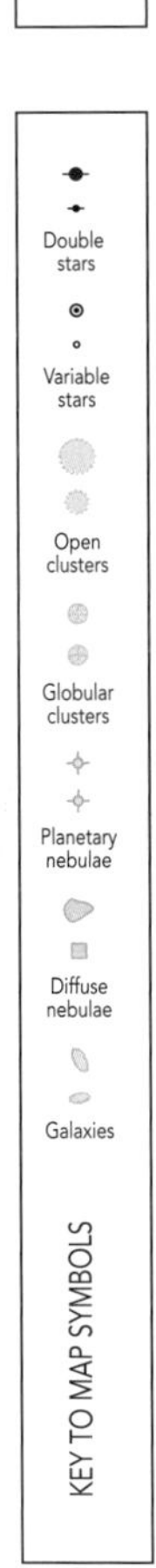
Double stars
Variable stars
Open clusters
Globular clusters
Planetary nebulae
Diffuse nebulae
Galaxies
KEY TO MAP SYMBOLS

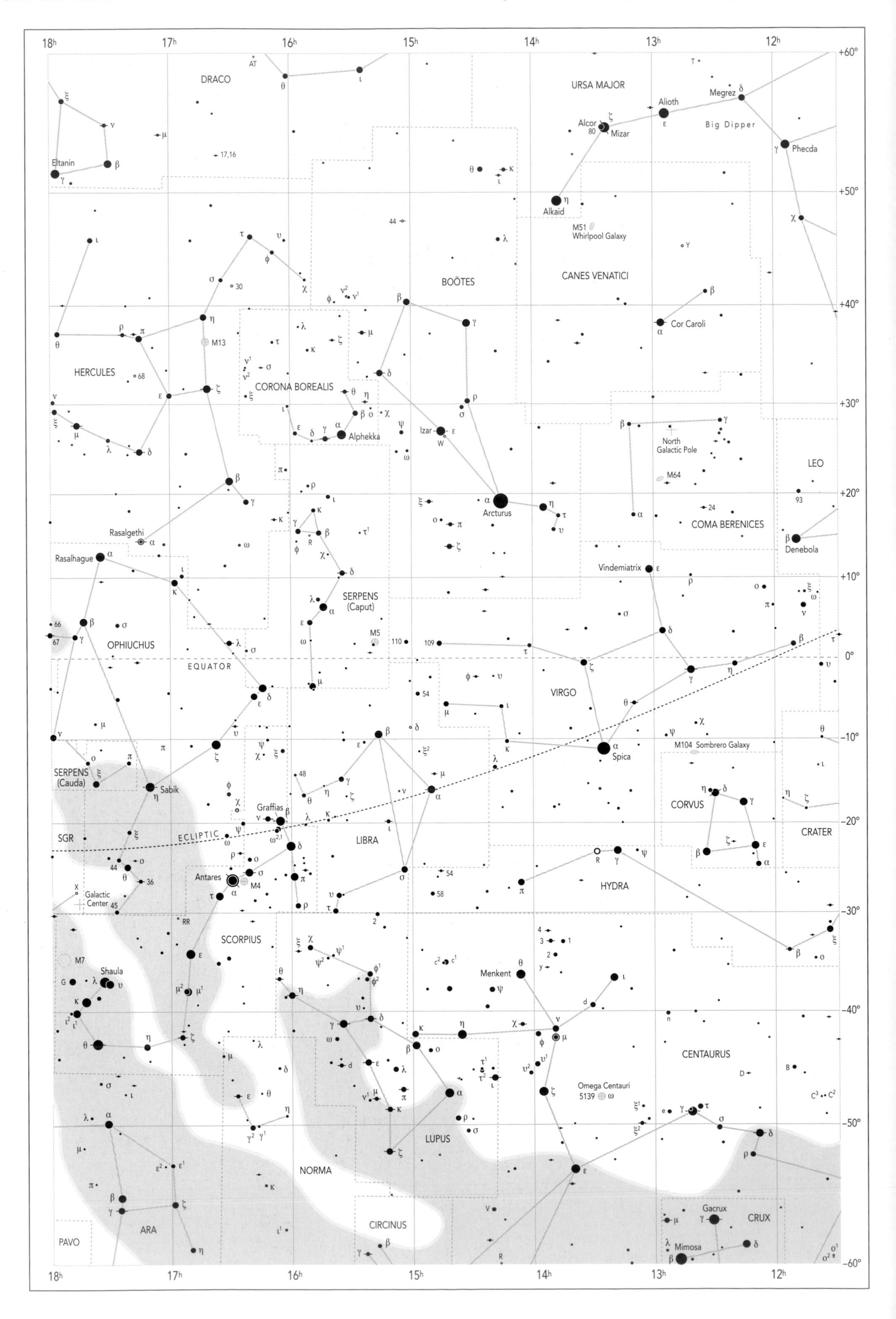
18h
17h
16h
15h
14h
13h
12h
+60°
+50°
+40°
+30°
+20°
+10°
0°
–10°
–20°
–30°
–40°
–50°
–60°
DRACO
Eltanin
URSA MAJOR
Alcor
Mizar
Alioth
Megrez
Big Dipper
Phecda
Alkaid
M51 Whirlpool Galaxy
CANES VENATICI
Cor Caroli
BOÖTES
HERCULES
M13
CORONA BOREALIS
Alphekka
Izar
North Galactic Pole
M64
LEO
Arcturus
COMA BERENICES
Denebola
Rasalgethi
Rasalhague
Vindemiatrix
SERPENS (Caput)
OPHIUCHUS
M5
EQUATOR
VIRGO
Spica
M104 Sombrero Galaxy
SERPENS (Cauda)
Sabik
CORVUS
CRATER
SGR
ECLIPTIC
Graffias
LIBRA
Antares
M4
HYDRA
Galactic Center
SCORPIUS
M7
Shaula
Menkent
CENTAURUS
Omega Centauri 5139
LUPUS
NORMA
ARA
CIRCINUS
Gacrux
CRUX
Mimosa
PAVO

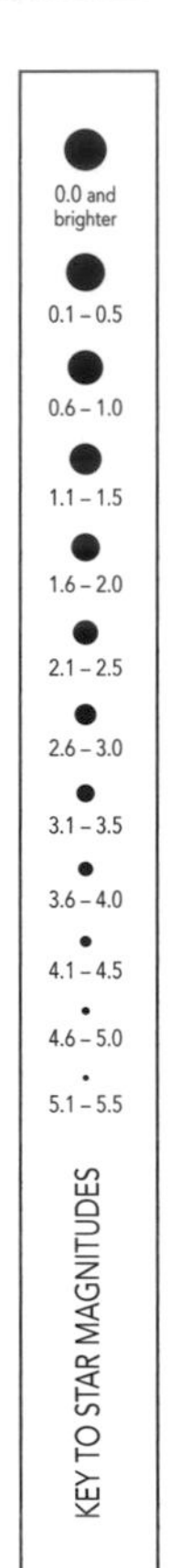
0.0 and brighter
0.1 – 0.5
0.6 – 1.0
1.1 – 1.5
1.6 – 2.0
2.1 – 2.5
2.6 – 3.0
3.1 – 3.5
3.6 – 4.0
4.1 – 4.5
4.6 – 5.0
5.1 – 5.5
KEY TO STAR MAGNITUDES

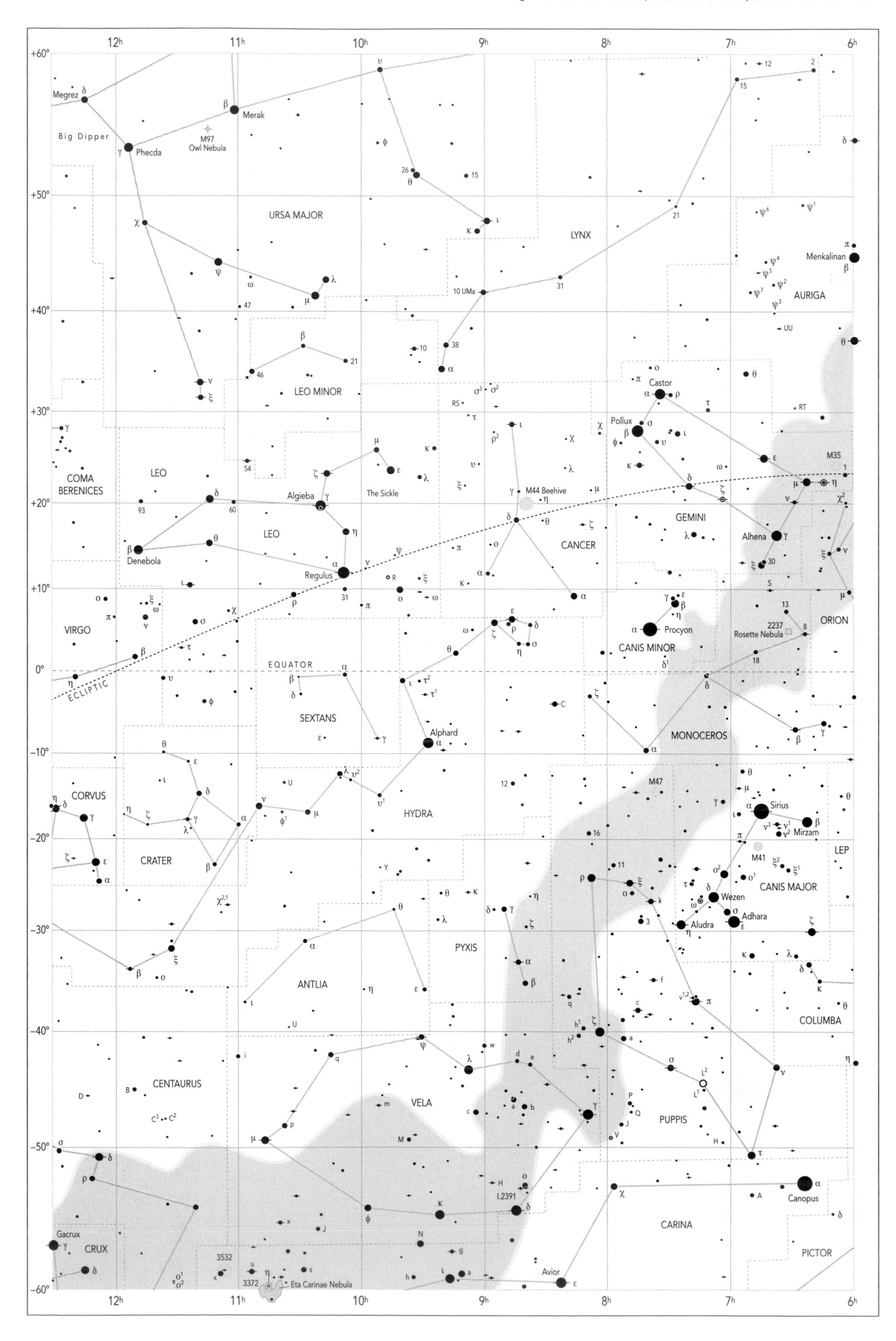
12h
11h
10h
9h
8h
7h
6h
+60°
+50°
+40°
+30°
+20°
+10°
0°
-10°
-20°
-30°
-40°
-50°
-60°
Megrez
Big Dipper
Phecda
Merak
M97 Owl Nebula
URSA MAJOR
LYNX
10 UMa
AURIGA
Menkalinan
LEO MINOR
Castor
Pollux
COMA BERENICES
LEO
Algieba
The Sickle
M44 Beehive
GEMINI
M35
Denebola
Regulus
CANCER
Alhena
ECLIPTIC
VIRGO
EQUATOR
Procyon
CANIS MINOR
Rosette Nebula
ORION
SEXTANS
Alphard
MONOCEROS
CORVUS
HYDRA
M47
Sirius
Mirzam
CRATER
M41
LEP
CANIS MAJOR
Wezen
Adhara
Aludra
PYXIS
ANTLIA
COLUMBA
CENTAURUS
VELA
PUPPIS
Canopus
CARINA
Gacrux
CRUX
Eta Carinae Nebula
Avior
PICTOR

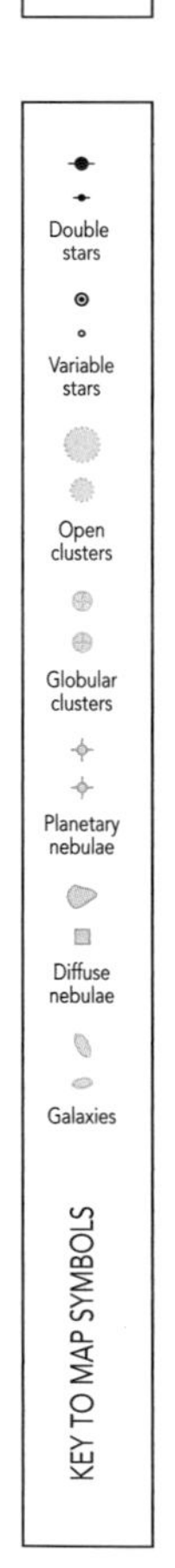
Double stars
Variable stars
Open clusters
Globular clusters
Planetary nebulae
Diffuse nebulae
Galaxies
KEY TO MAP SYMBOLS

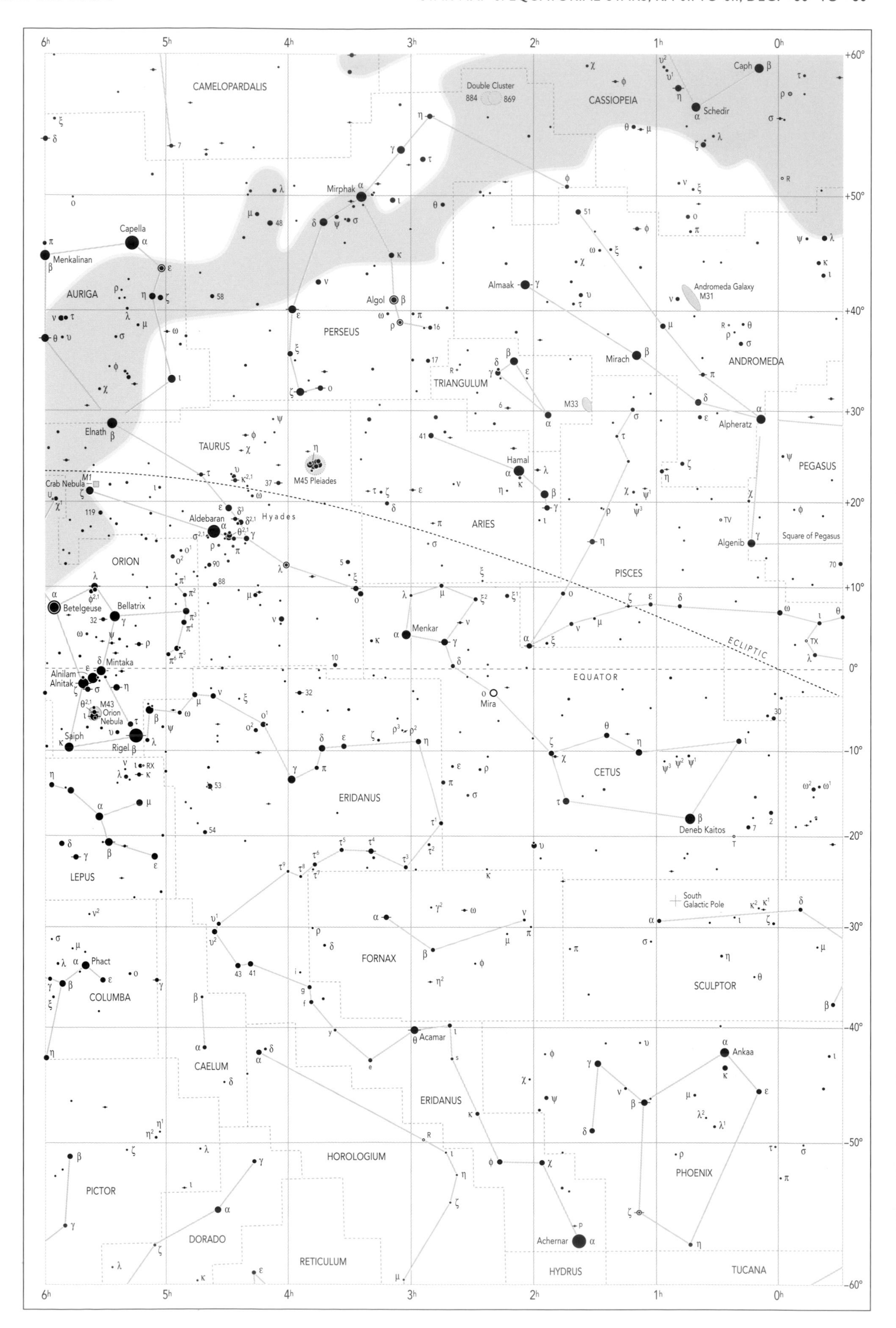
6h
5h
4h
3h
2h
1h
0h
+60°
+50°
+40°
+30°
+20°
+10°
0°
–10°
–20°
–30°
–40°
–50°
–60°
CAMELOPARDALIS
Double Cluster
884
869
CASSIOPEIA
Caph
Schedir
Mirphak
Capella
Menkalinan
AURIGA
Algol
Almaak
PERSEUS
Andromeda Galaxy
M31
Mirach
ANDROMEDA
TRIANGULUM
M33
Alpheratz
Elnath
TAURUS
M45 Pleiades
Hamal
PEGASUS
Crab Nebula
M1
Aldebaran
Hyades
ARIES
Algenib
Square of Pegasus
ORION
PISCES
Betelgeuse
Bellatrix
Menkar
ECLIPTIC
Mintaka
Alnilam
Alnitak
EQUATOR
Mira
M43
Orion
Nebula
Saiph
Rigel
CETUS
ERIDANUS
Deneb Kaitos
LEPUS
South
Galactic Pole
FORNAX
Phact
COLUMBA
SCULPTOR
Acamar
CAELUM
ERIDANUS
HOROLOGIUM
PHOENIX
PICTOR
Ankaa
DORADO
Achernar
RETICULUM
HYDRUS
TUCANA

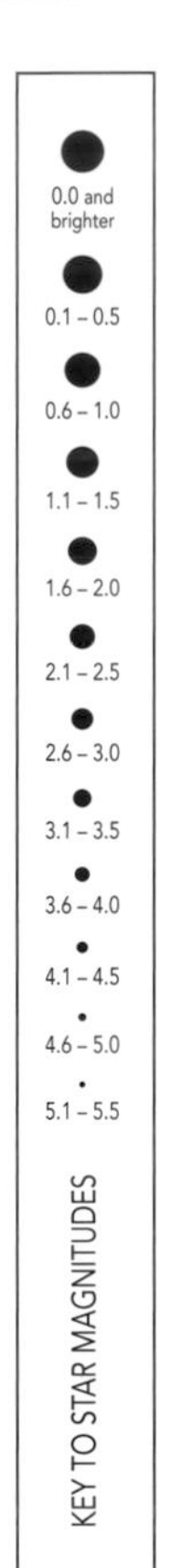
0.0 and brighter
0.1 – 0.5
0.6 – 1.0
1.1 – 1.5
1.6 – 2.0
2.1 – 2.5
2.6 – 3.0
3.1 – 3.5
3.6 – 4.0
4.1 – 4.5
4.6 – 5.0
5.1 – 5.5
KEY TO STAR MAGNITUDES

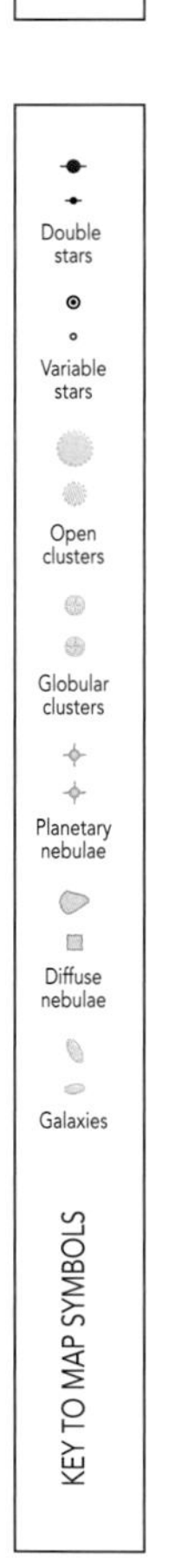
Double stars
Variable stars
Open clusters
Globular clusters
Planetary nebulae
Diffuse nebulae
Galaxies
KEY TO MAP SYMBOLS

STAR MAP 7: SOUTHERN STARS TO DEC. –30°

KEY TO STAR MAGNITUDES

0.0 and brighter
0.1 – 0.5
0.6 – 1.0
1.1 – 1.5
1.6 – 2.0
2.1 – 2.5
2.6 – 3.0
3.1 – 3.5
3.6 – 4.0
4.1 – 4.5
4.6 – 5.0
5.1 – 5.5

KEY TO MAP SYMBOLS

Double stars
Variable stars
Open clusters
Globular clusters
Planetary nebulae
Diffuse nebulae
Galaxies

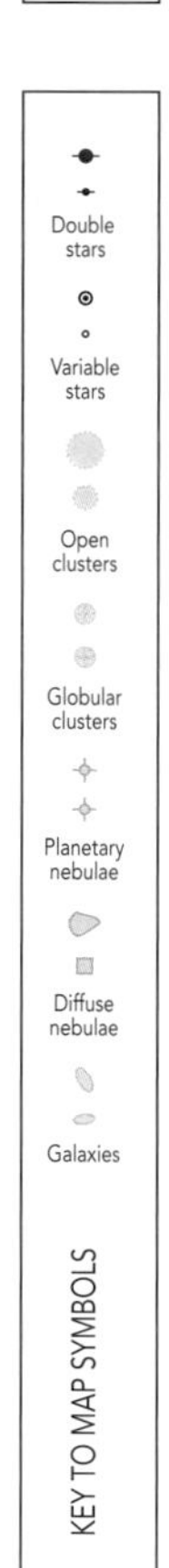
0.0 and brighter
0.1 – 0.5
0.6 – 1.0
1.1 – 1.5
1.6 – 2.0
2.1 – 2.5
2.6 – 3.0
3.1 – 3.5
3.6 – 4.0
4.1 – 4.5
4.6 – 5.0
5.1 – 5.5
KEY TO STAR MAGNITUDES
Double stars
Variable stars
Open clusters
Globular clusters
Planetary nebulae
Diffuse nebulae
Galaxies
KEY TO MAP SYMBOLS

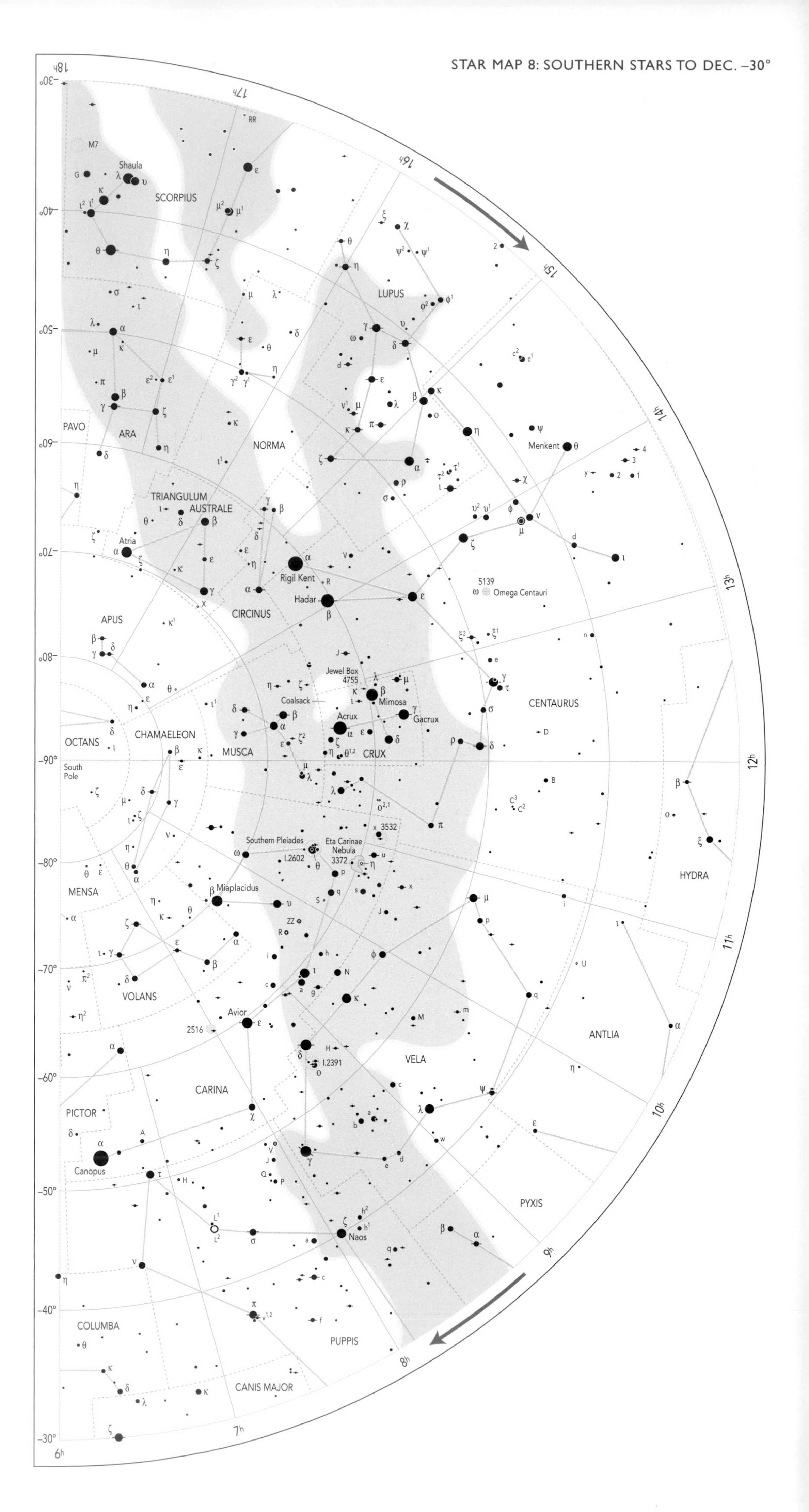
SCORPIUS
Shaula
M7
LUPUS
ARA
NORMA
PAVO
TRIANGULUM AUSTRALE
Atria
Rigil Kent
Hadar
CIRCINUS
APUS
Menkent
5139
Omega Centauri
CENTAURUS
Jewel Box 4755
Coalsack
Mimosa
Acrux
Gacrux
CRUX
MUSCA
CHAMAELEON
OCTANS
South Pole
HYDRA
Southern Pleiades
Eta Carinae Nebula 3372
I.2602
3532
Miaplacidus
MENSA
VOLANS
ANTLIA
Avior
2516
I.2391
VELA
CARINA
PICTOR
Canopus
PYXIS
Naos
COLUMBA
PUPPIS
CANIS MAJOR

CONSTELLATION	STAR MAP
Andromeda	1, 3, 6
Antlia	5
Apus	8
Aquarius	3
Aquila	3
Ara	3, 4, 8
Aries	6
Auriga	1, 2, 5, 6
Boötes	4
Caelum	6
Camelopardalis	1, 2
Cancer	5
Canes Venatici	4
Canis Major	5
Canis Minor	5
Capricornus	3
Carina	8
Cassiopeia	1
Centaurus	4, 5, 8
Cepheus	1
Cetus	6
Chamaeleon	8
Circinus	4, 8
Columba	5, 6
Coma Berenices	4
Corona Australis	3, 7
Corona Borealis	4
Corvus	4
Crater	5
Crux	8
Cygnus	3
Delphinus	3
Dorado	6, 7
Draco	3
Equuleus	3
Eridanus	6
Fornax	6
Gemini	5
Grus	3
Hercules	3, 4
Horologium	6, 7
Hydra	4, 5
Hydrus	7
Indus	7
Lacerta	1, 3
Leo	5
Leo Minor	5
Lepus	5, 6
Libra	4
Lupus	4, 8
Lynx	2, 5
Lyra	3
Mensa	7, 8
Microscopium	3, 7
Monoceros	5
Musca	8
Norma	8
Octans	7, 8
Ophiuchus	3, 4
Orion	5, 6
Pavo	3, 7, 8
Pegasus	3
Perseus	1, 6
Phoenix	3, 6
Pictor	5, 6, 7, 8
Pisces	3, 6
Piscis Austrinus	3
Puppis	5, 8
Pyxis	5
Reticulum	6, 7
Sagitta	3
Sagittarius	3, 4
Scorpius	4
Sculptor	3, 6
Scutum	3
Serpens	3, 4
Sextans	5
Taurus	6
Telescopium	3, 7
Triangulum	6
Triangulum Australe	8
Tucana	3, 7
Ursa Major	2, 4, 5
Ursa Minor	1, 2
Vela	5, 8
Virgo	4
Volans	8
Vulpecula	3

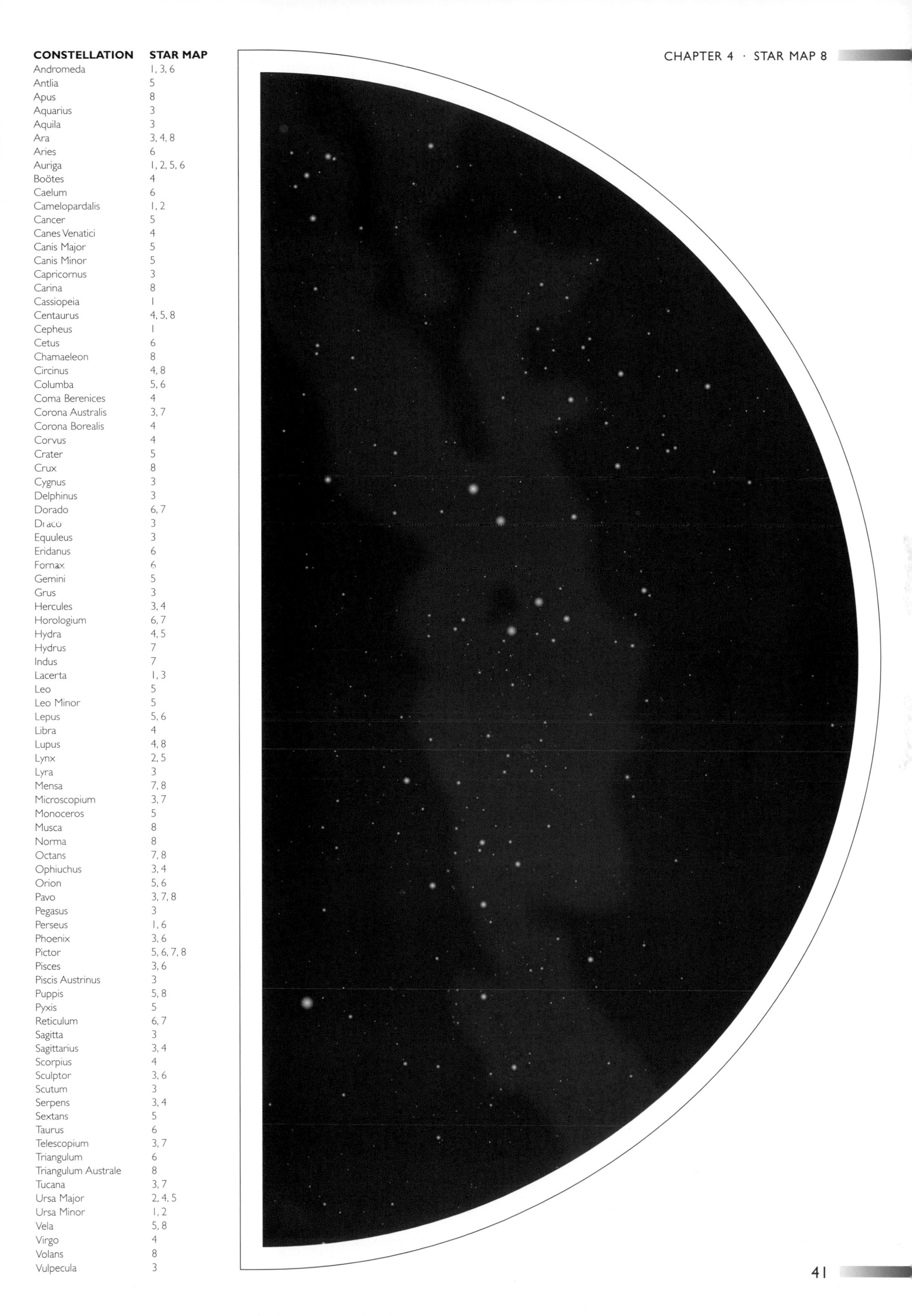

CHAPTER 5
THE MOON

To many amateur and indeed professional astronomers, the Moon is the villain of the piece. The period known as "the light of the Moon," when the Moon is high in the sky, renders observations of faint objects impossible, while "the dark of the Moon" is the joyous period when deep sky observing can take place. But to others, the Moon itself is a fascinating world in its own right, and far from being an unchanging object is full of surprises and hitherto unknown aspects.

In fact, if you want to impress someone with the view through your telescope, you can do no better than to show them the Moon at First Quarter. That familiar pale orb in the sky becomes transformed into a world in its own right, with peaks and valleys, plains and of course craters – thousands of craters, of all sizes. The Moon has been battered ever since its formation, 4.5 billion years ago. True, the Earth has been under similar attack by Solar System bodies of all sizes, but in our case most traces of the impacts have been obliterated by the action of water and wind, and also by tectonic activity and volcanism.

▲ *The Moon at its most spectacular, just after First Quarter.*

On the Moon there are no continental plates and erosion takes place much more slowly (over millions of years, rather than hundreds or thousands), so the features that you see now are the same as would have been seen at the time of our earliest human ancestors, a million or so years ago, or possibly even at the time of the dinosaurs, 65 million years ago. The asteroid impact that is thought to have wiped out the dinosaurs would have produced a major crater about 180 km across had it hit the Moon, but there it would be just one among many others.

Because the Moon is airless, not only do its features remain unchanged over the eons, but also the view with your telescope is virtually as good as if you were visiting it in a spaceship. Viewed with a magnification of 100, on a night when our own atmosphere is steady, you can imagine that you are 100 times closer to the Moon, and are getting the same view as the Apollo astronauts did when they were only 4000 km away from it instead of 400,000 km. The brilliance of the landscape can almost hurt your eyes, and the shadows are jet black. Although it is a wasteland, it is one that invites you to explore it, and you can do this from the comfort of your back garden.

There is something else to be said for the Moon, particularly for beginners. It provides an excellent test object on which you can hone your observing skills. Even if you can't find deep sky objects, you can hardly miss the Moon. If stars will not come into proper focus, the Moon will help you to see where you are going wrong, because there is no doubt when you are in focus. You can practise your drawing or photography on an object that is easy to track and well lit.

Movements of the Moon

As Earth's natural satellite, the Moon is always with us. Only for a few days each month, when it is more or less in line with the Sun, is it not visible at all. It orbits Earth every 27.3 days, so you might think that this is the interval between successive Full Moons. But during this time the Earth has continued in its orbit round the Sun, so the Moon must move a little more in its orbit to catch up. As a result, the interval between Full Moons is 29.4 days, a period known as a *lunation*.

We are so familiar with the face of the Moon in the sky that we never wonder why this is the case. If the Moon orbits us, surely we should see different sides of it? In fact, it has what is called *captured rotation*, which means that it always turns the same side toward the Earth. This is easy enough to grasp, but people often have a problem understanding that it really is rotating. The fact is that it rotates on its axis, as seen from the Sun, every 29.4 days. Imagine standing on a fairground roundabout, facing the center all the time, just as the near side of the Moon always faces Earth. For each revolution of the roundabout you have actually rotated once yourself, so someone standing outside would have seen all sides of you, while someone at the center of the roundabout would have seen only your face.

To extend the analogy, imagine that the only illumination is a floodlight some distance away from the roundabout. As you go round, it will sometimes be behind you, sometimes to your side and sometimes shining in your eyes. The same thing

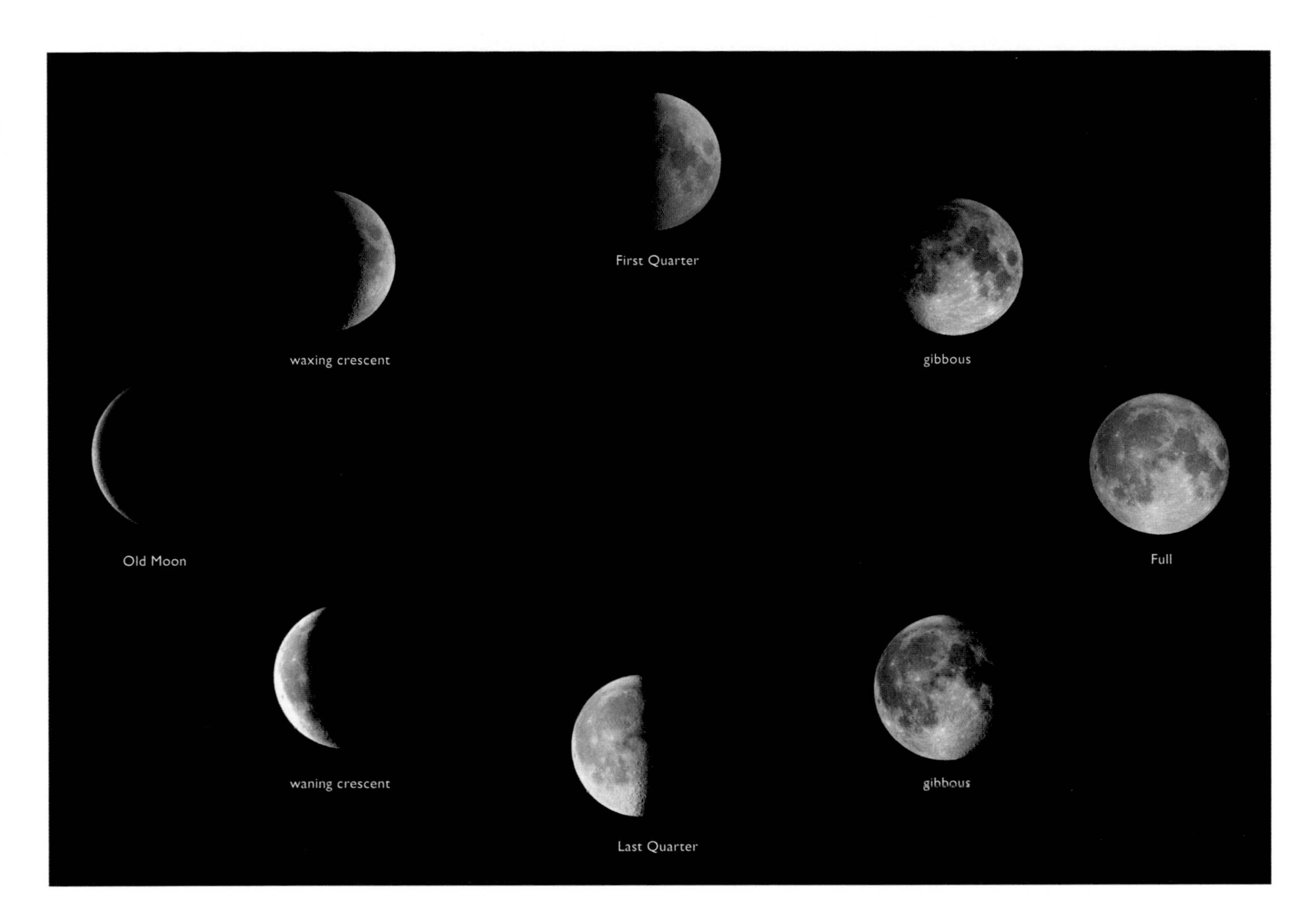

▲ *The phases of the Moon during one complete lunation.*

happens to the Moon, and this is what causes the *phases* – the changing angle of illumination from the Sun as the Moon goes round the Earth.

The lunation cycle begins at *New Moon*, when the Moon is between us and the Sun. Only rarely does it pass directly between the two, so that we see an eclipse (see page 68) – usually, it passes above or below the line of sight. We commonly refer to New Moon, however, as the thin crescent that appears in the evening sky just after sunset. Often you can see the rest of the Moon faintly illuminated by what is known as *Earthshine*. This is caused by light reflecting off the Earth, which is nearly full as seen from the Moon when the Moon is new to us. At this point in the cycle, the Moon sets shortly after the Sun.

Each day the Moon moves eastward in its orbit and is visible later and later in the evening as a larger and larger illuminated crescent. Seven days after New Moon it is at what is called *First Quarter* – because it is a quarter of the way around its orbit – and we see a half Moon. After this, its illuminated portion becomes larger and we call it *gibbous*. At this stage it is waxing, or growing in phase.

Seven days after First Quarter we get *Full Moon*, though for a day or two on either side of full it can look completely illuminated. At this point it is opposite the Sun in the sky, and therefore rises in the east at sunset. From then on it begins to wane, and the shadowed area starts to encroach on its landscape from the opposite edge. Each night it rises later and later until after another seven days or so it is at *Last Quarter*, when we again see a half Moon but with the opposite side illuminated from First Quarter. Last Quarter Moon is usually seen only in the morning sky, though at some times of the year it may rise well before midnight, particularly if you live at fairly high latitudes, such as in northern Europe.

After a few more days the Moon is a thin crescent again, this time rising just before the Sun in the east. It can be referred to as the *Old Moon* at this stage. It then lines up with the Sun and the cycle begins again.

People often refer to "the dark side of the Moon," meaning its far side. But all sides of the Moon are illuminated at some point during its orbit. At Full Moon the far side is indeed all dark, but at

◄ *Earthshine, sometimes called "the Old Moon in the New Moon's arms," is caused by sunlight reflecting off the Earth on to the unilluminated part of the Moon.*

New Moon, when it is between the Sun and the Earth, the far side is in full sunlight and it is the near side that is dark.

In fact, because the Moon does not orbit the Earth in a circle but in an ellipse, we can see a little more than half of its surface. Sometimes we see a little more of one edge, and at others we see more of the other edge. You can see this even with the naked eye if you are familiar with its surface. The dark area known as the Mare Crisium, near the western edge of the Moon as it is seen in the sky, is easily visible to most people. At times it is almost on the edge, while at others it is noticeably closer to the center. These variations in attitude are called *librations*, and lunar observers make use of them to see regions that are otherwise unseen.

The far side of the Moon has only been seen by Apollo astronauts and photographed by spacecraft.

Observing the Moon

Virtually any telescope will reveal details on the Moon. It does not take much magnification to show the craters and major features, which are just on the limit of visibility when the Moon is seen with the naked eye. Even binoculars show some detail, but will leave you wanting more power. The magnification limit is often set by our own atmosphere – on a night of bad seeing, a low power of 80 or so is the most you can use before the details become blurred. In these conditions, the surface is constantly in motion, as if seen at the bottom of a fast-moving stream. You get an impression of what is there, but the fine details are hard to make out. On a night of average or good seeing, however, you can increase the magnification to 150, 250 or perhaps even more.

▲ The Full Moon photographed on two different occasions reveals changes in tilt due to libration. In the top picture Plato is well away from the northern limb while Tycho appears quite near the southern limb. At bottom the situation is reversed.

Although high magnifications can give you the impression that you are flying just over the Moon's surface, the true size of lunar features visible with an amateur instrument is still quite large. With a 300 mm telescope, for example, observing under perfect seeing conditions, you could theoretically see features about 750 m or half a mile across. Even the Pentagon in Washington, D.C., would be too small to be distinguished if it were transferred to the Moon.

Some telescopes are equipped with a Moon filter, either green or neutral color. The purpose of this filter is simply to reduce the brightness of the Full Moon, which can be so bright as to leave you virtually blind when you step away from the eyepiece.

Because the Moon's phase is continually changing, the same feature seen on successive nights will appear different on each occasion. When it first appears in sunlight, on or near the shadow line or *terminator*, the feature is seen under a very low angle of illumination and as a result every slight bump or hollow in the surface casts a shadow. As the Sun rises over the feature, different details gradually come into sunlight, and when the Sun illuminates the feature from overhead all you can see are differences in brightness.

As the Moon takes almost a month to go through its cycle of phases, you might imagine that there will be little change in the illumination of the scene as you observe. However, variations in the lighting take place slowly but noticeably. If you decide to draw what you see – a traditional, satisfying, and quite easy means of recording the view – you will find that as soon as you have finished all but the quickest sketch, changes have started to occur, particularly if you are observing around the time of First Quarter.

Even if you are just Moongazing, one of the great delights of lunar observing is to watch the stately changes as the Sun rises over some great crater. To start with, only the highest peaks catch the sunlight. Soon, their lower flanks are visible, and before long the crater floor itself is crossed by the jagged shadows of the rim. After an hour or two, the whole crater is obvious. The next night you will find that the same crater is well away from the terminator, and is now blending into the mass of other craters.

Just as you learn the sky by finding a few signposts and building up your knowledge through the year, so you can learn the Moon bit by bit. It is unusual to be able to watch the sequence of phases night after night, either because of the weather or for social reasons, but you can return month after month on clear nights and pick up where you left off.

Names on the Moon

The lunar features were named in the early part of the telescopic era. The first observers did not realize what sort of a world they were looking at, and they imagined that the dark areas might be seas. These were given names using the common language of science at the time, Latin. The Latin name for sea is *mare*, pronounced "mah-ray," the plural being *maria*, pronounced "mah-ree-ah." So we have Mare Tranquillitatis, the Sea of Tranquillity, and so on. One suspects, however, that romance took over from reason when Mare Nectaris (Sea of Nectars) and Lacus Mortis (Lake of Death) received their names. The area where the Ranger 7 spacecraft crash-landed in 1964, sending back the first close-ups of the lunar surface, was subsequently called Mare Cognitum, meaning "Sea that has become known." We now know that the lunar seas are huge basins left after the impact of giant asteroids, which accounts for the roughly circular appearance of many of them.

The craters have traditionally been named after astronomers and other worthies, mostly historical ones. Indeed, in many cases a prominent lunar feature is our main reminder of someone who has since sunk into obscurity.

Compass points on the Moon all changed in the 1960s when the first lunar missions took place. Prior to that time, the west side of the Moon was taken to mean the west as seen from Earth, so the Mare Orientale, for example, was on the Moon's eastern edge – hence its name. But to a lunar explorer it would be to the west, and we now refer to lunar directions as if we were seeing the surface like a map. For the same reason, lunar maps are now shown with north at the top, although astronomical telescopes invariably invert the view. For observers in the southern hemisphere, however, an official map of the Moon is now the same way up as they see it through a telescope.

The maps on the subsequent pages mark in red the locations where various spacecraft, both manned and unmanned, have either landed or crashed.

SOME LUNAR FEATURES AND THEIR TRANSLATIONS	
Sinus Amoris	Bay of Love
Mare Anguis	Sea of Serpents
Mare Australis	Southern Sea
Mare Crisium	Sea of Crises
Palus Epidemiarum	Marsh of Diseases
Mare Cognitum	Known Sea
Mare Fecunditatis	Sea of Fertility
Mare Frigoris	Sea of Cold
Mare Humboldtianum	Humboldt's Sea
Mare Humorum	Sea of Moisture
Mare Imbrium	Sea of Rains
Mars Insularum	Sea of Islands
Sinus Iridum	Bay of Rainbows
Mare Marginis	Border Sea
Sinus Medii	Central Bay
Lacus Mortis	Lake of Death
Palus Nebularum	Marsh of Mists
Mare Nectaris	Sea of Nectar
Mare Nubium	Sea of Clouds
Mare Orientale	Eastern Sea
Oceanus Procellarum	Ocean of Storms
Palus Putredinis	Marsh of Decay
Sinus Roris	Bay of Dews
Mare Serenitatis	Sea of Serenity
Sinus Aestuum	Sea of Seething
Mare Smythii	Smyth's Sea
Palus Somni	Marsh of Sleep
Mare Spumans	Foaming Sea
Mare Tranquillitatis	Sea of Tranquillity
Mare Undarum	Sea of Waves
Mare Vaporum	Sea of Vapors

► *The craters Aristarchus (left) and Herodotus as seen from the Apollo 15 spacecraft in low lunar orbit.*

Northeast quadrant

Though it is small, everyone's eyes are attracted to the Mare Crisium when the Moon is an evening crescent. On the floor of the Mare Crisium are two craters, Picard and Peirce, which look small at first glance. However, they are respectively 34 km and 19 km in diameter, about the diameter of major cities on Earth. To the west of the Mare Crisium is a bright ray crater, Proclus, about the same size as Picard but much more prominent. Ray craters are surrounded by bright streaks of material thrown out when they formed as a result of an impact, and the craters themselves are the brightest features on the Moon. Proclus is unusual in that the rays are asymmetric, leaving the gray plain of Palus Somni untouched. The impacting body must have come in at an oblique angle, throwing debris ahead of its track. Although oblique impacts still result in circular craters, the splash marks that we see as rays betray their direction.

The famous Mare Tranquillitatis, site of the first lunar landing, lies to the west of the Palus Somni. Its floor is crossed by the two obvious fault lines of the Cauchy Rupes (wall) and Rima (fissure). Similar features lie west of Mare Tranquillitatis, namely the Rima Hyginus, also known as the Hyginus Rille. A line of craters extends along this fault line, showing where massive ground collapses took place. Though the vast majority of the Moon's craters are caused by impacts, these are examples of volcanic features.

▲ Mare Crisium, with the bright ray crater Proclus to its left (west) and the diamond-shaped Palus Somni.

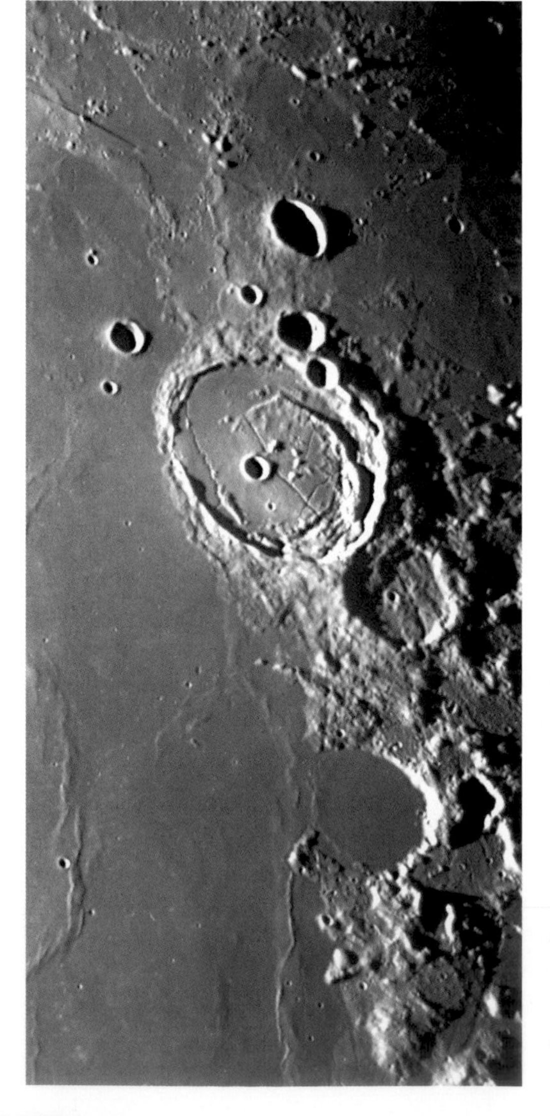

◄ The complex interior of Posidonius contains some fascinating details. Nearby is a wrinkle ridge known as Dorsa Smirnov.

Northwest of Mare Tranquillitatis lies Mare Serenitatis, flanked by the eroded crater Posidonius. This comes into the category of walled plain, as it has been worn down by the impact of countless tiny meteoric bodies over eons of time and now consists of a circular wall with a flat center. Features known as wrinkle ridges cover the floor of Mare Serenitatis, and are particularly obvious under low illumination. Perhaps these helped to give the impression that the Moon's dark areas are seas, with oddly stationary waves.

The remainder of this quadrant is lighter in color, with many craters and walled plains which you can explore over a period of time.

▼ The Hyginus Rille at center with the crater Triesnecker and the Triesnecker rilles to its south.

N
80°
70°
60°
50°
40°
30°
20°
10°
0°
0° E
Meton
Bond, W.
Archytas
Protagoras
F R I G O R I S
Galle
Aristoteles
Vallis Alpes
MONTES ALPES
Trouvelot
Egede
Mitchell
Baily
Endymion
MARE HUMBOLDTIANUM
Atlas
Hercules
Bürg
LACUS MORTIS
Eudoxus
Lamech
Mason
Plana
Grove
Alexander
Cassini
Callippus
MONTES CAUCASUS
LACUS SOMNIORUM
Theaetetus
Rimae Daniell
Daniell
Luther
PALUS NEBULARUM
Posidonius
Rima Bond
Bond, G.
Autolycus
MARE
Kirchhoff
Mons Hadley
PR. FRESNEL
Linné
Chacornac
Le Monnier
LUNA 21
MONTES TAURUS
Römer
SERENITATIS
Bessel
Conon
Deseilligny
Littrow
APOLLO 17
MONTES HAEMUS
Sulpicius Gallus
Mons Argaeus
Maraldi
Tacquet
Menelaus
PR. ARCHERUSIA
Dawes
Vitruvius
Plinius
MARE VAPORUM
Manilius
Jansen
Ross
Maclear
Boscovich
RANGER 6
Julius Caesar
Sosigenes
MARE
Sinas
Rupes Cauchy
Rima Cauchy
Cauchy
Ukert
Hyginus
Rima Hyginus
Rima Ariadaeus
Ariadaeus
Rimae Triesnecker
Silberschlag
Whewell
Arago
Murchison
Cayley
Manners
Triesnecker
Agrippa
TRANQUILLITATIS
Chladni
De Morgan
Dionysius
Dembowski
RANGER 8
Ritter
Blagg
Godin
Sabine
Schmidt
SURVEYOR 5
Rhaeticus
Rima Hypatia
APOLLO 11
Moltke
Theon Senior
Lade
Rima Oppolzer
Réaumur
Delambre
Theon Junior
Pickering
Horrocks
Saunder
Hipparchus
Taylor
Hypatia
Alfraganus
Torricelli
Gyldén
Müller
Hind
Zöllner
Halley
Andĕl
APOLLO 16
Isidorus
Capella
Gutenberg
Goclenius
FECUNDITATIS
Langrenus
La Pérouse
Kästner
Gilbert
Maclaurin
Webb
LUNA 16
MARE
Messier
Lubbock
Censorinus
Secchi
Maskelyne
LUNAS 18 AND 20
Apollonius
Dubiago
MARE SPUMANS
Schubert
MARE SMYTHII
MARE UNDARUM
Neper
Firmicius
Taruntius
Auzout
Da Vinci
Condorcet
Hansen
MARE CRISIUM
Lick
Glaisher
LUNA 24
PR. AGARUM
Picard
Yerkes
Proclus
Lyell
Franz
PALUS SOMNI
Peirce
LUNA 15
Alhazen
MARE MARGINIS
MARE ANGUIS
Tisserand
Macrobius
Eimmart
Plutarch
Delmotte
Cleomedes
Seneca
Tralles
Newcomb
Burckhardt
Hahn
Berosus
Geminus
Bernouilli
Berzelius
Maury
Franklin
Messala
Gauss
Hooke
Cepheus
Shuckburgh
Oersted
Chevallier
Carrington
Schumacher
Mercurius
Zeno
De La Rue
Strabo
Thales
Schwabe
Democritus
Gärtner
Kane
Mayer, Christian
Sheepshanks
Arnold
Moigno
Neison
Peters
Cusanus
Petermann
Fuctemon
Scoresby
Main
Gioja
Challis
Barrow

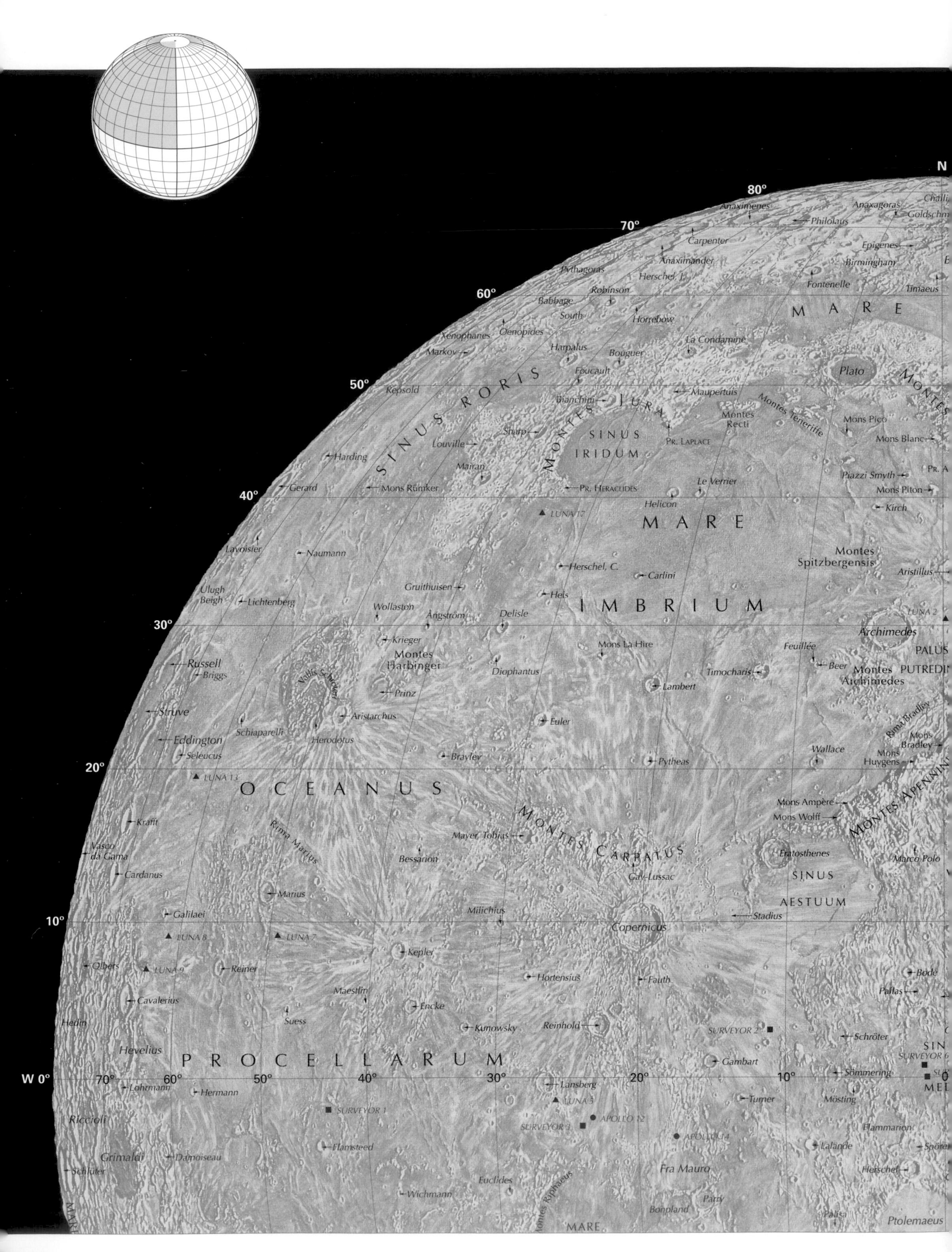
N
80°
70°
60°
50°
40°
30°
20°
10°
W 0°
70°
60°
50°
40°
30°
20°
10°
Anaximenes
Anaxagoras
Philolaus
Goldschm
Carpenter
Epigenes
Anaximander
Birmingham
Pythagoras
Herschel, J.
Fontenelle
Timaeus
Robinson
Babbage
South
Horrebow
M A R E
Xenophanes
Oenopides
La Condamine
Harpalus
Markov
Bouguer
Plato
Foucault
S I N U S R O R I S
Repsold
Maupertuis
Bianchini
M O N T E S J U R A
Montes Recti
Montes Teneriffe
Sharp
S I N U S I R I D U M
Mons Pico
Louville
Pr. Laplace
Mons Blanc
Harding
Mairan
Piazzi Smyth
Gerard
Mons Rümker
Pr. Heraclides
Le Verrier
Mons Piton
Helicon
Kirch
Luna 17
M A R E
Lavoisier
Naumann
Montes Spitzbergensis
Herschel, C.
Carlini
Aristillus
Ulugh Beigh
Lichtenberg
Gruithuisen
Heis
I M B R I U M
Wollaston
Ångström
Delisle
Luna 2
Archimedes
Krieger
Mons La Hire
Feuillée
Palus
Russell
Montes Harbinger
Beer
Montes Archimedes
Putredi
Briggs
Vallis Schröteri
Diophantus
Timocharis
Lambert
Prinz
Struve
Aristarchus
Rima Bradley
Euler
Schiaparelli
Mons Bradley
Eddington
Herodotus
Wallace
Seleucus
Brayley
Pytheas
Mons Huygens
Luna 13
O C E A N U S
Montes Apenninus
Mons Ampère
Mons Wolff
Krafft
Rima Marius
M O N T E S C A R P A T U S
Mayer, Tobias
Vasco da Gama
Bessarion
Eratosthenes
Marco Polo
Cardanus
Gay-Lussac
S I N U S
Marius
A E S T U U M
Milichius
Galilaei
Stadius
Copernicus
Luna 8
Luna 7
Kepler
Olbers
Luna 9
Reiner
Bode
Hortensius
Fauth
Pallas
Cavalerius
Maestlin
Encke
Hedin
Suess
Kunowsky
Reinhold
Surveyor 2
Schröter
Hevelius
P R O C E L L A R U M
Gambart
Surveyor 6
Sömmering
Lohrmann
Hermann
Lansberg
Turner
Mösting
Luna 5
Surveyor 1
Apollo 12
Riccioli
Surveyor 3
Flammarion
Apollo 14
Flamsteed
Lalande
Grimaldi
Damoiseau
Fra Mauro
Herschel
Schlüter
Euclides
Montes Riphaeus
Wichmann
Parry
Bonpland
Palisa
Ptolemaeus
MARE

Northwest quadrant

Many lunar observers find this the most fascinating section of the Moon, as it includes some spectacular features. Much of it is covered by sea, either the vast Oceanus Procellarum or the separate and dramatic basin of the Mare Imbrium. These plains are punctuated by very striking individual craters which make the area much easier to navigate than some other parts of the Moon.

The Mare Imbrium resulted from an enormous impact comparatively late in the Moon's cratering history, though at a time when the Solar System was still young. As you gaze at it through your telescope, you can see that the lava flows from the impact have flooded earlier craters, such as Stadius at its southern edge, and the beautiful curved bay of Sinus Iridum. In fact, the impact took place some 3900 million years ago, at a time when the first primitive single-celled life forms were starting to emerge on Earth. Subsequently, a comparatively small number of impacts created such magnificent features as Copernicus and Kepler. The fact that the lava plains of Imbrium have comparatively few craters shows that the size and number of impacts must have dropped off quite sharply since the formation of Mare Imbrium itself. (Counts of craters are a standard means of assessing the age of particular landscapes on planets or their satellites.)

Copernicus is surrounded by bright rays and strings of craterlets that were formed by the debris, known as *ejecta*, that was thrown from the crater during the massive explosion that created it about 800 million years ago. It is a classic large lunar crater, with a central peak and slumped walls. Small craters, just a few kilometers across, remain as perfect bowl shapes, but the larger ones are much flatter in profile. The central peak was formed as a result of the rebound of the lunar surface, much like the central drop that you see in high-speed photos of a drop falling into milk. The lunar surface is not strong enough to support high crater walls so they slump into numerous terraces.

▲ *The magnificent Mare Imbrium. Plato is at top right, with the Sinus Iridum to its left. The crater at the bottom of the terminator is Eratosthenes.*

To the west of Copernicus lies Aristarchus, a ray crater that is the brightest feature on the Moon. Adjacent to it, issuing from a point near the older crater Herodotus, is Vallis Schröteri, a channel left after the flow of lava. This valley winds for 150 km across the surface of the Oceanus Procellarum.

Other notable features in this quadrant are the distinctive gray-floored walled plain Plato; the 130 km Vallis Alpes fault which cuts through the lunar Alps or Montes Alpes; and the lone peaks of Pico and Piton, standing 2800 m and 2300 m above the Mare Imbrium. These peaks are the first to catch the sunlight, shining like stars in the darkness.

◀ *Copernicus is surrounded by the debris that was thrown out when it was formed. To its east are the ghostly remains of a flooded crater called Stadius.*

▶ *Though it looks like a gash in the Alps caused by the formation of the Mare Imbrium, the Alpine Valley (or Vallis Alpes) is a rift valley like the much larger example in East Africa. A rille within it is visible under good conditions with larger telescopes.*

▶ *The flooded walled plain Plato. At first glance the craters on its floor are easily missed.*

The pictures on these pages have north at the top, and most were taken through amateur telescopes.

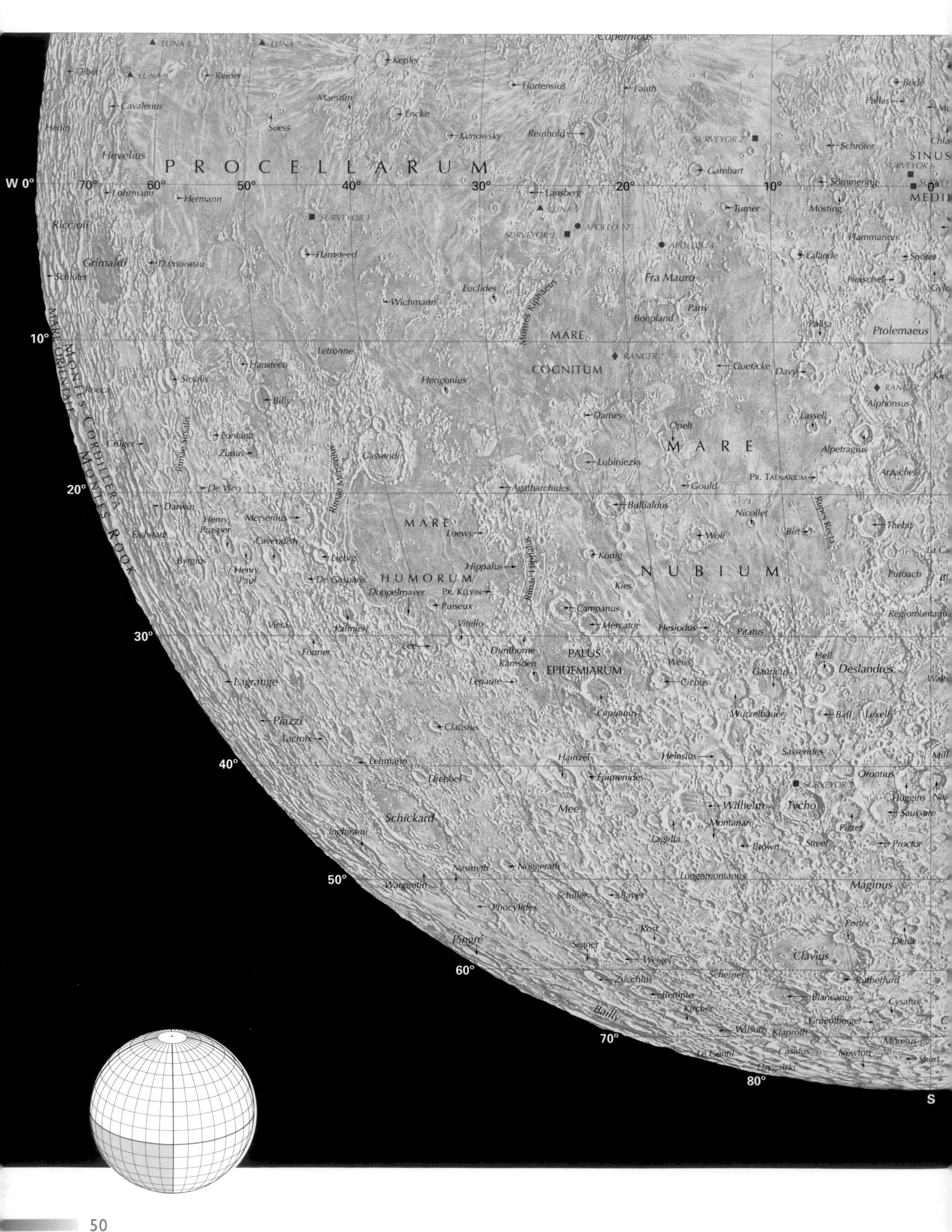

LUNA 8
LUNA 7
Kepler
Olbers
LUNA 9
Reiner
Hortensius
Fauth
Bode
Pallas
Cavalerius
Maestlin
Encke
Hedin
Suess
Kunowsky
Reinhold
SURVEYOR 2
Schröter
Hevelius
SINUS
SURVEYOR 6
PROCELLARUM
Gambart
W 0°
70°
60°
50°
40°
30°
20°
10°
0°
Lohrmann
Hermann
Lansberg
Sömmering
LUNA 5
Turner
Mösting
Riccioli
SURVEYOR 1
APOLLO 12
SURVEYOR 3
APOLLO 14
Flammarion
Grimaldi
Damoiseau
Flamsteed
Lalande
Spörer
Schlüter
Fra Mauro
Herschel
Euclides
Montes Riphaeus
Wichmann
Parry
Bonpland
Palisa
Ptolemaeus
MARE ORIENTALE
MONTES CORDILLERA
MONTES ROOK
MARE
10°
Letronne
RANGER 7
Hansteen
COGNITUM
Guericke
Davy
Sirsalis
Rocca
Herigonius
RANGER 9
Billy
Alphonsus
Darney
Lassell
Opelt
Fontana
Crüger
Rimae Sirsalis
Zupus
Gassendi
Rimae Mersenius
MARE
Lubiniezky
Alpetragius
Arzachel
PR. TAENARIUM
20°
De Vico
Agatharchides
Gould
Darwin
Bullialdus
Rupes Recta
Nicollet
Mersenius
Henry Prosper
MARE
Thebit
Eichstadt
Loewy
Birt
Cavendish
Wolf
Byrgius
Liebig
Rimae Hippalus
König
Henry Paul
Hippalus
NUBIUM
Purbach
De Gasparis
HUMORUM
Doppelmayer
PR. KELVIN
Kies
Puiseux
Campanus
Vieta
Palmieri
Vitello
Mercator
Hesiodus
Pitatus
30°
Fourier
Lee
Dunthorne
PALUS
Ramsden
EPIDEMIARUM
Weiss
Hell
Deslandres
Lagrange
Lepaute
Cichus
Gauricus
Capuanus
Wurzelbauer
Ball
Lexell
Piazzi
Clausius
Lacroix
Sasserides
Lehmann
Hainzel
Heinsius
40°
Epimenides
Orontius
Drebbel
SURVEYOR 7
Huggins
Wilhelm
Tycho
Mee
Saussure
Schickard
Montanari
Pictet
Inghirami
Lagalla
Brown
Street
Proctor
Nasmyth
Nöggerath
50°
Wargentin
Longomontanus
Maginus
Schiller
Bayer
Phocylides
Porter
Rost
Pingré
Segner
Deluc
Weigel
Clavius
60°
Scheiner
Zucchius
Rutherfurd
Bettinus
Blancanus
Kircher
Bailly
Cysatus
Gruemberger
70°
Wilson
Klaproth
Moretus
Le Gentil
Casatus
Newton
Short
Drygalski
80°
S

Southwest quadrant

This quadrant is full of interesting landmarks. The first notable feature to be seen each lunation is the north–south line of craters Ptolemaeus, Alphonsus, and Arzachel, sometimes referred to as the Lunar Snowman from its sequence of decreasing size. Ptolemaeus has a dark, flat floor with a prominent crater, Ammonius, just 9 km across. Just to the west of Arzachel in the Mare Nubium is one of the most appealing features on the Moon, the Rupes Recta or Straight Wall. This is a scarp fault some 110 km long, and it appears as a dark line around First Quarter or as a bright line around Last Quarter. Despite its name it is not particularly steep, with a slope of around 40°.

South of the Mare Nubium lies a large area of lunar highland. Here can be found one of the most curious sights on the Moon, the crater Tycho. With a diameter of 85 km, it is just a few kilometers smaller than Copernicus and as a recent crater it has many similarities (though it is in the lighter highlands rather than in a sea). When it is close to the terminator it is noticeably crisper than most of the surrounding craters, and it has the same imposing ramparts and central peak as Copernicus. But while most craters merge into the background as the Sun climbs higher over them, Tycho gets ever brighter. By Full Moon it is the most eye-catching feature on the whole Moon, as it is a brilliant ray crater.

Tycho's rays cross a large area of the Moon, though there is a darker area devoid of rays immediately surrounding the actual crater. Tycho looks just like a globe with lines of longitude radiating away from the pole, and people sometimes believe that it must be the south pole – though of course real poles do not have convenient grid lines on them!

▲ *The Mare Humorum with its great shoreline crater Gassendi. On its opposite shore is the fragmented wall of Doppelmayer.*

Closer to the actual south pole than Tycho lies one of the Moon's largest craters, Clavius, with a diameter of 225 km. Only Bailly, near the limb to the west, is larger. Such a crater is so large that if you were standing in the center, you would not see the outer walls – thanks in part to the Moon's smaller diameter than the Earth, and therefore closer horizons. Clavius contains a distinctive arc of craters of decreasing size.

West of the Mare Nubium lies a roughly circular sea called Mare Humorum. At the sea's north side lies a fascinating crater, Gassendi, which has a floor covered with rilles or fault lines.

At the far western edge of the Moon is a dark oval walled plain, Grimaldi, which looks like a miniature version of Mare Crisium on the opposite limb. Grimaldi is the darkest feature on the Moon.

▲ *Clavius (lower center) and Tycho (top left) are two of the best-known craters of the highlands area. Tycho is said to be a mere 50 million years old.*

► *Ptolemaeus (top), Alphonsus, and Arzachel are an easily recognized group of craters close to the center of the Moon's disk. Each crater has its own individual characteristics.*

Southeast quadrant

The first glimpse of this quadrant comes when the Moon is a thin crescent and two great craters, Langrenus and Petavius, are on the terminator. Langrenus, about 130 km across, is on the shore of the Mare Fecunditatis, which contains the unusual pair of ray craters Messier and Messier A. The body that created these must have come in at a very low angle indeed, so that it skipped across the surface like a stone skimming across water, creating the two craters. Messier has two rays extending north–south on either side of it, while Messier A has two almost parallel rays in line with the direction of motion of the impacting body.

The adjacent sea, Mare Nectaris, is notable for several large craters on its shores. Fracastorius has been breached by the sea itself, while the trio of Theophilus, Cyrillus, and Catharina line the northwestern shore. An arc-shaped escarpment known as the Altai Scarp or Rupes Altai surrounds the sea on the eastern side and is very conspicuous as a bright line before First Quarter or by its shadows before Last Quarter. Its southern end is marked by the crater Piccolomini.

Much of the quadrant is occupied by light-colored highland, with numerous overlapping craters. Among the most prominent craters are the ancient ruined feature Janssen, which is one of the largest features on the Moon; Stöfler, with its floor darker than the surroundings; and Maurolycus.

▼ The crater trio Theophilus (top), Cyrillus (center), and Catharina (bottom) lie close to the Altai Scarp, which under waning illumination, as seen here, appears dark.

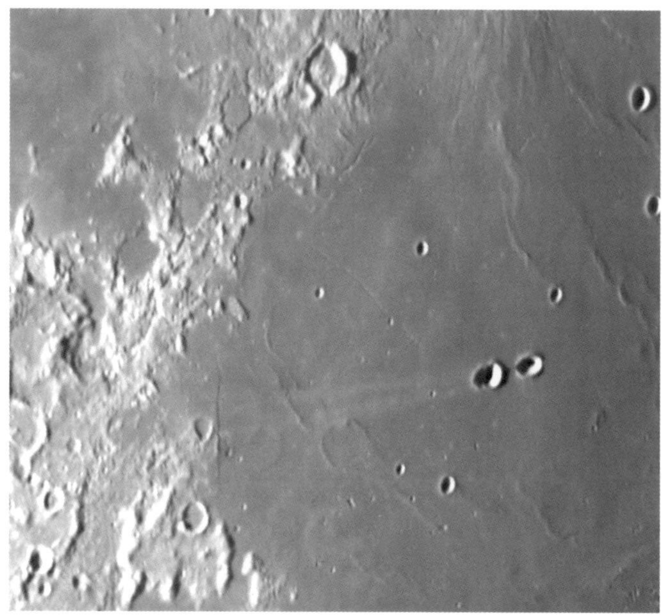

▲ The fascinating crater pair Messier (right) and Messier A in the Mare Fecunditatis. They were formed by an oblique impact. The rays from Messier are only visible under high illumination.

◄ Janssen occupies the center of this picture, but it lies under the more recent craters Fabricius and Metius to the north and Lockyer to the south.

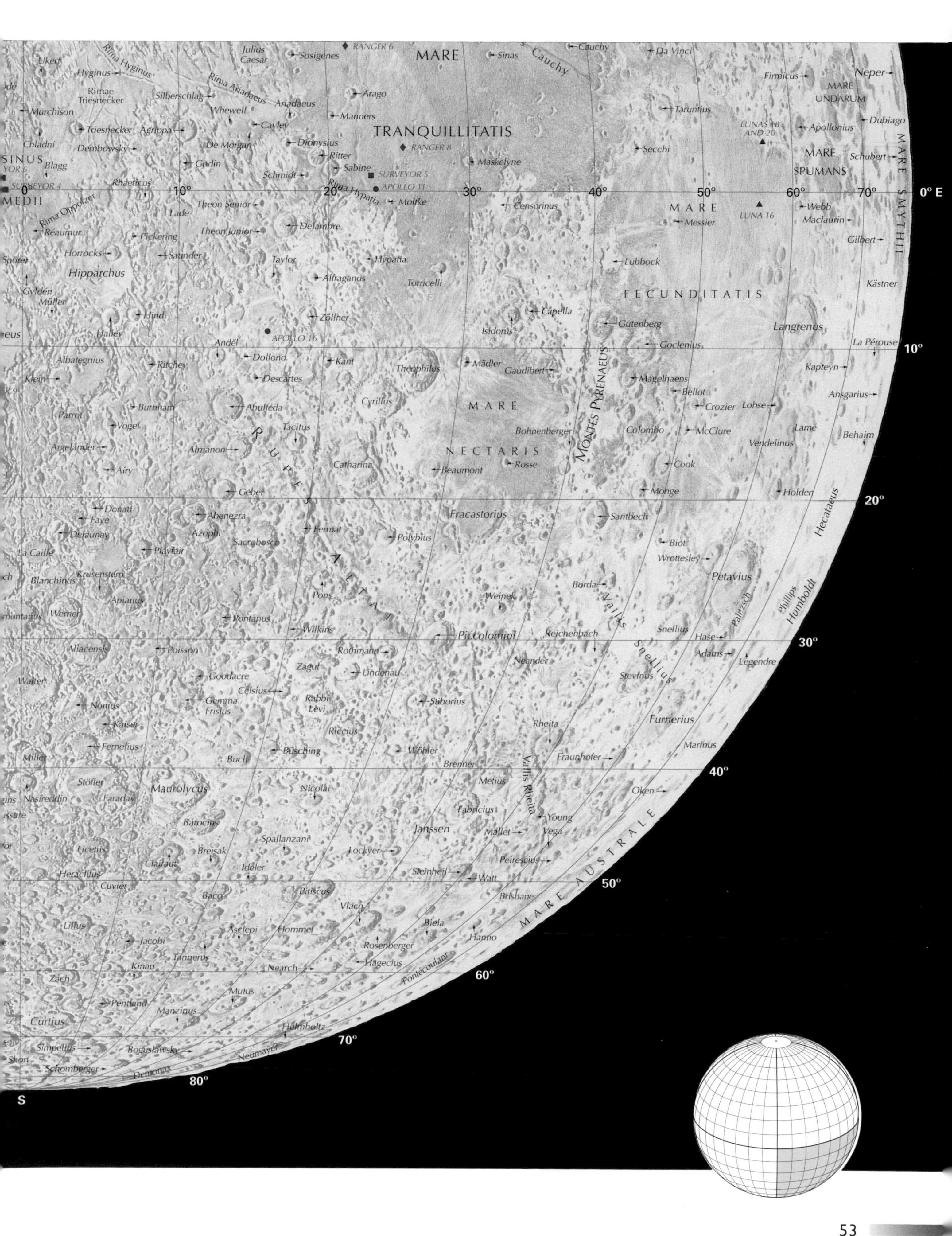
Ukert
Rima Hyginus
Hyginus
Julius Caesar
Sosigenes
RANGER 6
MARE
Sinas
Cauchy
Cauchy
Da Vinci
Rimae Triesnecker
Silberschlag
Rima Ariadaeus
Ariadaeus
Arago
Firmicus
Neper
MARE UNDARUM
Murchison
Whewell
Manners
Taruntius
Triesnecker
Agrippa
Cayley
TRANQUILLITATIS
LUNAS 18 AND 20
Apollonius
Dubiago
Chladni
De Morgan
Dionysius
RANGER 8
Secchi
Dembowsky
Ritter
MARE SPUMANS
Schubert
SINUS
Blagg
Godin
Maskelyne
Sabine
SURVEYOR 5
Schmidt
Rhaeticus
APOLLO 11
SURVEYOR 4
MEDII
Rima Hypatia
Moltke
0°
10°
20°
30°
40°
50°
60°
70°
0° E
Rima Oppolzer
Theon Senior
Censorinus
MARE
Messier
LUNA 16
Webb
Maclaurin
MARE SMYTHII
Lade
Réaumur
Delambre
Theon Junior
Pickering
Gilbert
Spörer
Horrocks
Saunder
Taylor
Hypatia
Lubbock
Hipparchus
Alfraganus
Torricelli
Kästner
Gyldén
Müller
FECUNDITATIS
Capella
Hind
Zöllner
Gutenberg
Langrenus
Halley
Isidorus
APOLLO 16
Anděl
Goclenius
La Pérouse
10°
Dollond
Albategnius
Kant
Ritchey
Mädler
Theophilus
Kapteyn
Gaudibert
Descartes
Klein
Magelhaens
Bellot
Ansgarius
Abulfeda
Burnham
Cyrillus
Crozier
Lohse
MARE
MONTES PYRENAEUS
Parrot
Vogel
Tacitus
Colombo
McClure
Lamé
Behaim
Bohnenberger
Vendelinus
Argelander
Almanon
NECTARIS
Airy
RUPES ALTAI
Catharina
Beaumont
Rosse
Cook
Geber
Monge
Holden
20°
Hecataeus
Donati
Faye
Abenezra
Fracastorius
Santbech
Azophi
Delaunay
Fermat
Polybius
Sacrobosco
Biot
La Caille
Playfair
Wrottesley
Krusenstern
Blanchinus
Petavius
Apianus
Pons
Borda
Weinek
Vallis Snellius
Phillips
Humboldt
Werner
Palitzsch
Pontanus
Wilkins
Snellius
Piccolomini
Reichenbach
Hase
Aliacensis
30°
Poisson
Rothmann
Adams
Neander
Legendre
Zagut
Goodacre
Lindenau
Stevinus
Walter
Celsius
Rabbi Levi
Gemma Frisius
Nonius
Stiborius
Kaiser
Riccius
Rheita
Furnerius
Fernelius
Marinus
Büsching
Wöhler
Miller
Buch
Fraunhofer
Brenner
Vallis Rheita
Stöfler
40°
Maurolycus
Metius
Nicolai
Oken
Nasireddin
Faraday
Fabricius
Young
Barocius
Janssen
Vega
Mallet
Spallanzani
Licetus
Breisak
Lockyer
Peirescius
Clairaut
Heraclitus
Ideler
Steinheil
Watt
50°
Cuvier
Baco
Pitiscus
Brisbane
MARE AUSTRALE
Vlacq
Lilius
Biela
Asclepi
Hommel
Hanno
Jacobi
Rosenberger
Tannerus
Hagecius
Kinau
Nearch
Pontécoulant
60°
Zach
Mutus
Pentland
Manzinus
Curtius
Helmholtz
70°
Simpelius
Boguslawsky
Neumayer
Short
Schomberger
Demonax
80°
S

CHAPTER 6
THE SOLAR SYSTEM

The Sun

The Sun is one of the easiest astronomical objects to observe, but it can also be the most difficult. It is easy because it is available at social hours, is unaffected by light pollution, and requires only a small telescope for good observations. It is difficult because it is blindingly bright, and there is an ever-present risk that you may unwittingly be blinded permanently.

People sometimes say, "But I often have to look directly at the Sun when I'm driving at sunset, so why is it so dangerous?" There are several reasons. One is that when you see the Sun in the sky, you take only a brief glimpse before your reflexes make you look away, or your view dances about so that no one part of your retina receives the heat of the Sun for very long. Another is that when the Sun is low in the sky its intensity is considerably reduced. A third is that when viewed with the naked eye, the Sun's image falls on only a small part of the retina. When seen through a telescope, however, its heat is spread over a wide area and the blood vessels cannot conduct the heat away quickly enough to avoid permanent damage.

There is a tale that early observers, such as Galileo, ruined their eyesight by staring at the Sun through a telescope. But Galileo had more sense, and he was the first to realize that the simplest way to view the Sun is to project its image through the telescope on to a piece of paper or card held behind the eyepiece. This is exactly what we do today, though there are also alternative high-tech methods.

Three steps to solar observing, here using a 110 mm reflector.

1 Cap the finder and point the telescope at the Sun using its shadow as a guide.

2 Hold a card about 30 cm behind the eyepiece to project the Sun's image.

3 Focus the image.

Recall those experiments with a magnifying glass focusing the Sun's image to set fire to paper and you will appreciate that there is a risk to your telescope as well as to your eyesight. Telescopes with plastic tubes or eyepiece barrels are totally unsuited to solar observation, and should not be used at all. Some instruction manuals for more solidly built instruments also advise against projecting the Sun's image to avoid damage to the optics. The danger is to the eyepiece, which could become overheated and either crack or suffer damage to the resin that cements the glass elements to each other. You can reduce the risk of damage by not leaving the telescope pointing at the Sun for long periods and by maybe using basic eyepiece types, such as Huygenian eyepieces, for solar projection. Before you project the Sun's image, make sure that the cap is on the finder telescope so that no one tries to look through it, and also to avoid damage to its eyepiece.

Unlike most other astronomical objects, where light is at a premium, in the case of the Sun the problem is that there is too much of it. This is why many telescopes have a tube cover with a small aperture that has its own separate cap. These are intended for use when observing the Sun, to reduce the amount of light reaching the eyepiece. In the case of reflectors the cap is off-center, so that the aperture is unobstructed by the secondary mirror and the spider assembly that holds it in place.

Point the telescope at the Sun by using its shadow as a guide. Use a low-power eyepiece and hold a piece of white card about 30 cm behind the eyepiece, in place of your eye. You should see a bright circle of light on the card, probably out of focus. Refocus until the disk of light is sharp – this means focusing until the image size is at its smallest. This should be the disk of the Sun.

Though there may be genuine sunspots, you may also see the out-of-focus image of any dust on the eyepiece. You can confirm which is which by rotating the eyepiece in the focusing mount. Any dust and eyepiece defects will rotate, but the Sun's image will not.

To get a larger though dimmer image, move the card farther away from the eyepiece and refocus. If the Sun is dimmed by haze or you have a very small telescope, you may need to move the card closer to the eyepiece to get a brighter, smaller view. You can experiment with higher-power eyepieces, which will probably give a view of only part of the Sun. A homemade card sunshield on the telescope or the screen will help to keep direct sunlight off the projected image. You can get large, dramatic views by projecting the image into a darkened room, but the image will move quickly and the setup is difficult to maintain for any length of time.

Solar filters

These days, dense full-aperture solar filters are readily available which allow you to observe the Sun directly through the telescope. Do not be tempted to use other dense materials as filters because they may not have the required infrared absorption (see page 68). Solar filters must fit snugly over the aperture of the telescope – that is, over the top of the tube – where they are subject to normal sunlight rather than the focused beam. The inexpensive filters are made of thin film on which is deposited a metallized coating, and are perfectly safe as long as you follow elementary safety rules. There must be no chink of unexposed aperture, nor damage to the filter (such as by creasing, though some wrinkling is normal), and you must make sure that there is no danger of them falling off or perhaps being removed as a prank.

At one time, all small telescopes were supplied with Sun filters that screw into an eyepiece. These are universally outlawed by the astronomical community because they are known to heat up and crack, with devastating consequences. You may still come across them on telescopes that were made in the past, but never use them.

More advanced filters are available that transmit only a narrow set of wavelengths, usually the light from hydrogen atoms (hydrogen alpha). The advantage of these filters is that they will reveal the prominences on the Sun that are normally only visible during a solar eclipse, and they also show considerable detail on the Sun's surface that is not visible in white light. However, the price tag for the filter alone is about the same as that for a medium-sized telescope. Do not confuse these very narrowband hydrogen-alpha solar filters with much cheaper broadband deep-red filters designed for astrophotography.

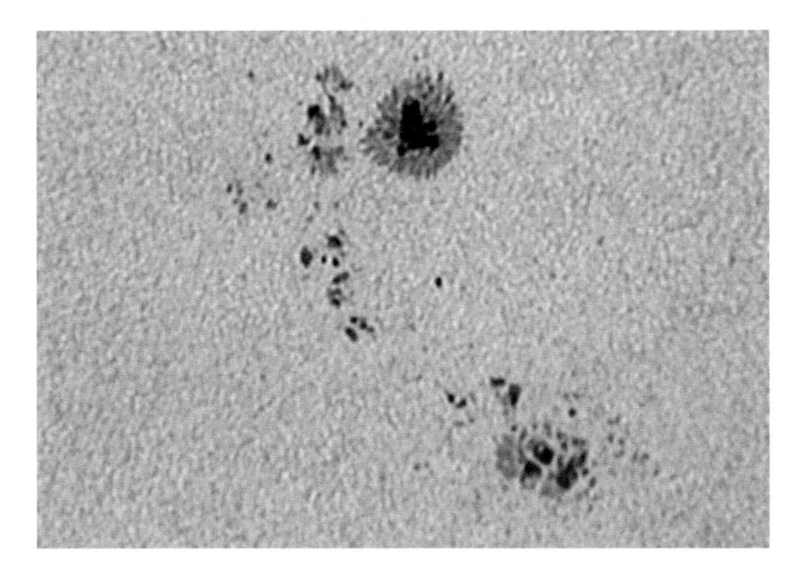

▲ *A typical small sunspot group shown in close-up. The image was taken on May 26, 2004, when the group was close to the center of the Sun's disk. The rice-grain texture of the Sun's surface is evident.*

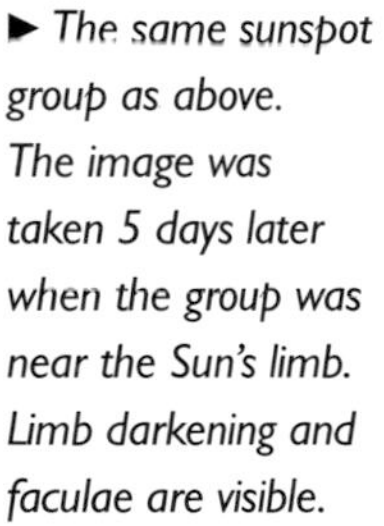

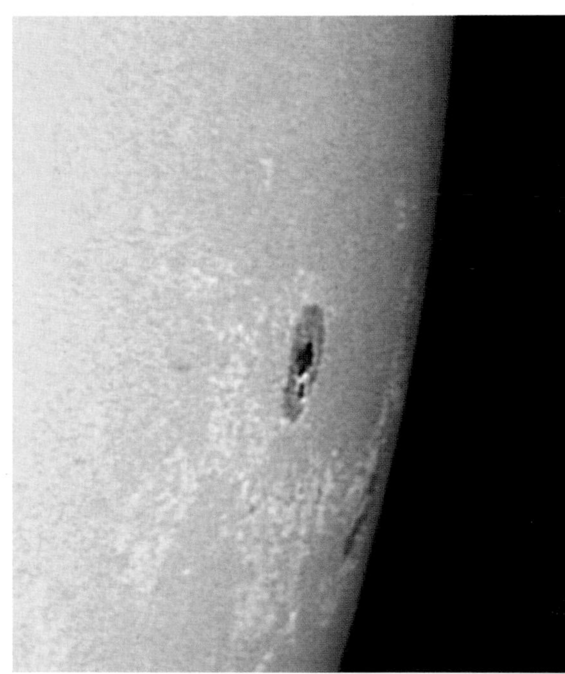

► *The same sunspot group as above. The image was taken 5 days later when the group was near the Sun's limb. Limb darkening and faculae are visible.*

What you can see

The Sun is a gaseous body, and what we see as the surface is actually a layer known as the *photosphere* where the maximum light is emitted. Above it, but normally too faint to be seen, is a layer known as the *chromosphere*. People sometimes think that haze around the Sun is the chromosphere, but this is actually haze caused by our own atmosphere.

Some features of the Sun's disk are a result of the photosphere being a gaseous layer rather than a solid surface. The higher regions of the photosphere are cooler, and so shine less brightly than the deeper regions. At the center of the disk we see directly into the lowest layers, but nearer the edge, or limb, we look through a greater thickness of gas and see the higher, dimmer layers. This means that the edge of the Sun appears slightly darker, an effect known as *limb darkening*. Hot clouds of gas that hang higher in the photosphere are known as *faculae*, and they are usually visible near the limb against the limb darkening.

Everyone has heard of solar flares, but these are usually wrongly identified with the prominences that are seen like red flames at the edge of the Sun during total solar eclipses (see page 68) or when using hydrogen-alpha filters. Flares are outbursts of solar energy, and on rare occasions they are visible as white spots or bars within sunspots. They last only a few minutes, and can give rise to streams of solar particles.

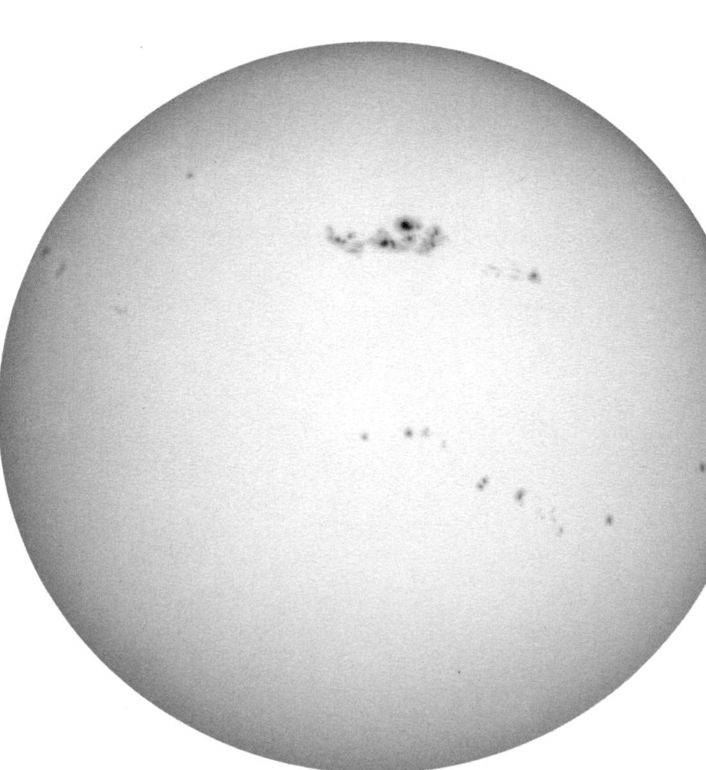

▲ *This projected image of the Sun was photographed with a digital camera and shows it in its natural color, white. It was taken in 2001 when there were large numbers of sunspots visible.*

Sunspots

The most obvious features on the Sun are sunspots. These dark markings on the solar disk are usually present. They are areas where the Sun's light output is restricted to a certain extent by

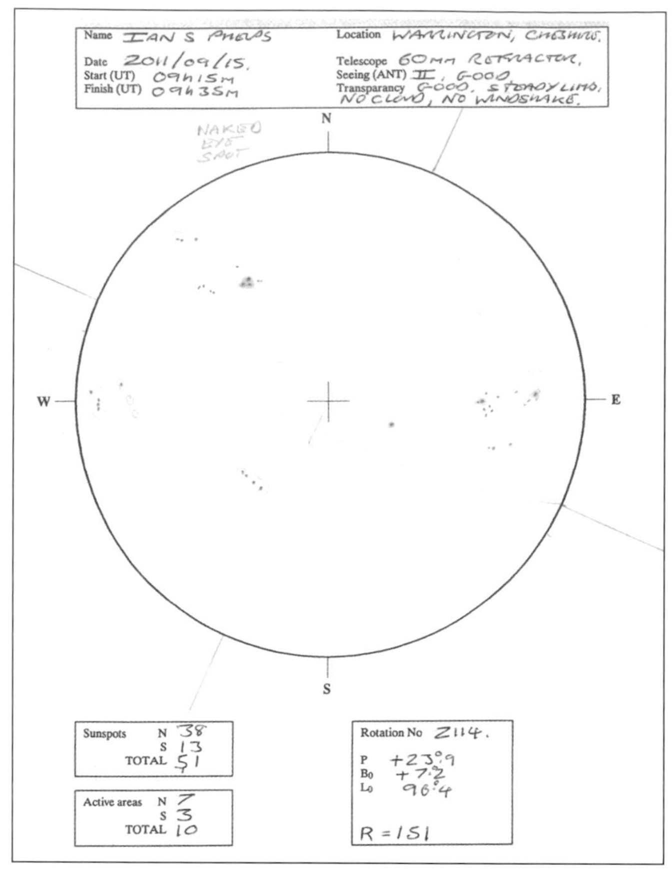

▲ *A detailed solar drawing made using a 60 mm refractor. The Sun's equator and other particulars have been added afterward, and can be found from published tables or online, having established the east–west line from the drift of a sunspot.*

magnetic fields at that point in the photosphere. In fact sunspots do emit a considerable amount of light, but in comparison with the rest of the Sun's surface they appear dark. An individual spot has a dark center, known as the *umbra*, surrounded by a lighter region, the *penumbra*. If the seeing is good, you will notice that the whole solar surface has a granular structure, sometimes called a rice-grain structure, which consists of convection cells where the hot gas rises and cooler gas falls.

▼► *Specialized solar telescopes, such as the Coronado PST, are available with hydrogen-alpha filters that show any prominences visible as well as additional details on the Sun.*

Sunspots also have this granular appearance, but the grains are often elongated in the direction of the center of the spot.

Spots often appear in pairs or in groups. Small ones may last for only an hour or two, while big ones – usually massive complex groups – can be larger in extent than the planet Jupiter and may last for weeks. A group of sunspots is known as an *active region* or active area.

The Sun rotates from east to west every 27 days or so as seen from the Earth's point of view, so spots appear to move from one side of the solar disk to the other, changing day by day. A major active region may last for more than one solar rotation. Such regions are often a source of other solar features, notably prominences and flares, and when a major active region is on the Sun, even if it has just appeared on its eastern limb, there is an increased chance of an aurora (see page 71).

The numbers of sunspots vary over a roughly 11-year cycle. At solar maximum the Sun is usually covered with spots, while at minimum there may be very few, or even none at all. Some maxima are stronger than others, and in recent years the number of sunspots at maximum has been noticeably greater than it was 100 years ago. As a new cycle begins, spots appear at high latitudes on the solar disk, though they rarely appear above a solar latitude of 35°. As the cycle continues spots are more common closer and closer to the equator, and as the cycle ends high-latitude spots from the new cycle may overlap with those at the equator from the old cycle.

The simplest way to record sunspots is to project the Sun's image on to a standard 15 cm prepared circle on your projection sheet. You can then draw in the spots and faculae (as outlines) directly, and can keep track of the movement of spots across the disk. Make a note of the date and time. At some point switch off the telescope drive if you have one, so that the image drifts across the paper. Mark the direction of drift of a sunspot so that you can get the orientation of the image the same from day to day – the drift is toward the west.

To help determine the position of the Sun's axis and for much more information on solar observing, go to the website www.petermeadows.com/.

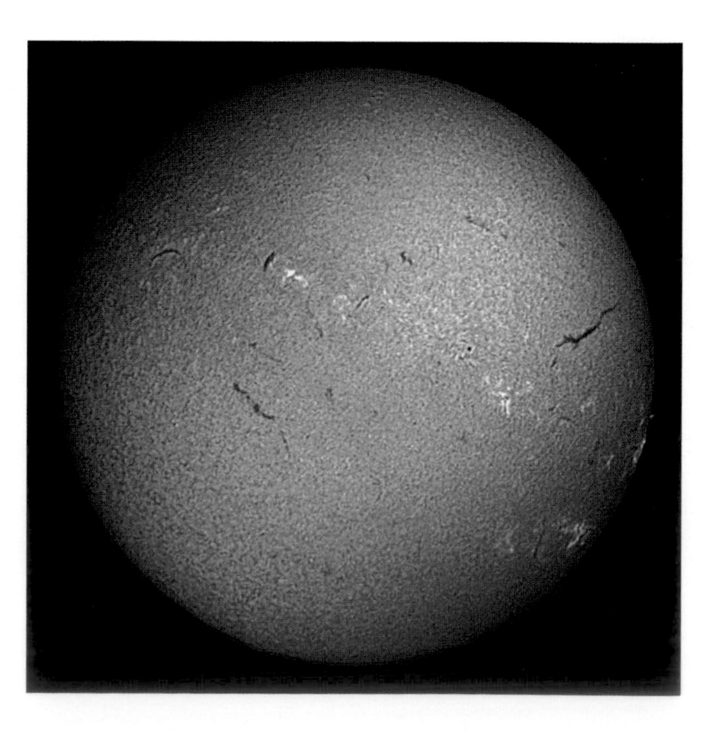

Mercury

Mercury is not hard to find, and it is surprising how many otherwise experienced observers say that they have never seen it. However, you do need to look at just the right time and in the right place. Because the planet stays so close to the Sun and moves so quickly around it, there is a period of only about 10 days at each elongation (see page 12) when it is far enough from the Sun to be seen. If the ecliptic is at a shallow angle to the horizon, it is effectively impossible to see even then.

When you do find Mercury, you may see a tiny white or pinkish disk through a telescope. The disk is small, and you will need at least a 75 mm telescope to see anything. Only elusive and vague markings are ever visible. Mercury goes through a cycle of phases similar to that of the Moon. When it is at full or crescent phase, however, it is too close to the Sun to be observed easily, and only a limited range on either side of half phase is visible. Occasionally Mercury passes across the face of the Sun, in what is called a *transit*. It is then visible as a tiny black disk against the Sun, smaller than many sunspots.

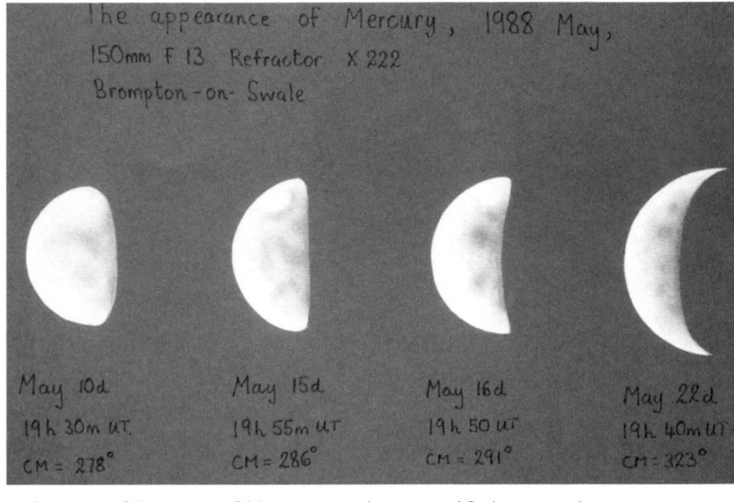

▲ *A series of drawings of Mercury made over a 12-day period. They show the changes in appearance of the planet and the markings visible on the tiny disk.*

As a planet, Mercury is a rocky, airless world, similar in general appearance to the Moon except that it lacks the vast lava seas. The Messenger spacecraft is on a mission to photograph the planet and has sent back detailed images of its surface, but no spacecraft has ever landed there. Data for all the planets are to be found on page 126.

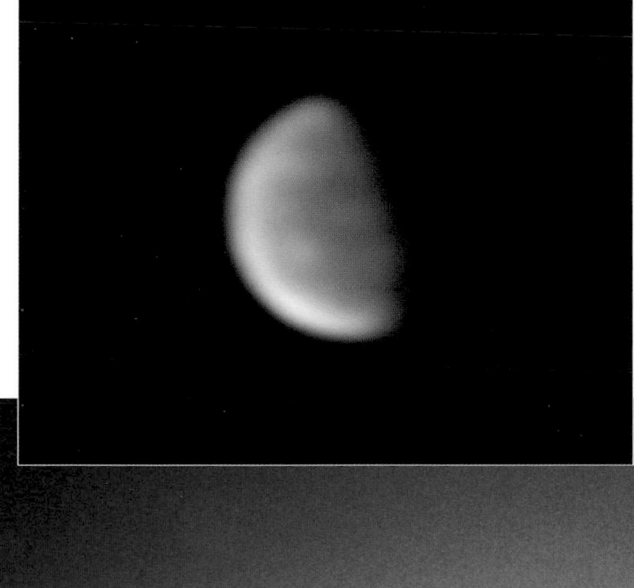

◄ *A photograph of Mercury made using a webcam and a standard 200 mm Schmidt-Cassegrain telescope. Markings are evident, as is the reddish color of the planet.*

▼ *Mercury is only ever visible for a few days at a time low down in the twilight sky (here, just above the trees at center).*

Venus

If Mercury is elusive, its companion planet, Venus, is the most highly visible. On those occasions when it is at its brightest in the evening sky, people who rarely give the skies a second glance will notice it as "the evening star." If this happens around Christmas time, people may ask if it is a second Star of Bethlehem; at other times of year its appearance spawns a flurry of UFO sightings. Venus is the brightest object in the sky after the Sun and Moon, and is far brighter than any other planet or star, at magnitude –4.

Like Mercury, Venus goes through a cycle of phases. It starts an evening apparition with an almost full disk when it is on the far side of the Sun. Eight months later, when it is at its greatest eastern elongation, it is at half phase. After only another two months, just before it crosses between us and the Sun, it is a thin crescent. On very rare occasions Venus transits across the disk of the Sun, but normally it passes above or below because its orbit is inclined at a different angle to that of the Earth.

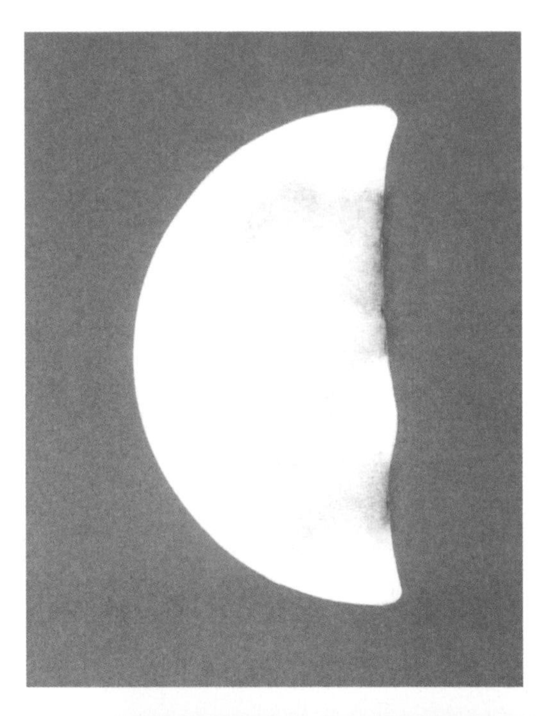

◄ A drawing of Venus at just over half phase, with the terminator line noticeably crooked. The drawing was made about 45 minutes before sunset.

▼ A wider range of phases can be observed on Venus than on Mercury because it moves farther from the Sun. These photographs were taken using a 280 mm telescope and a webcam.

The reason why Venus is so bright is that it is completely covered in cloud, which reflects sunlight very effectively. Its dense atmosphere is composed of carbon dioxide, and the pressure at its surface is about 90 times that of our own atmosphere. The surface temperature is very high – around 480°C – and this, combined with a sulfuric-acid cloud layer, makes conditions on Venus very hostile. The few spacecraft, all Soviet, to have landed there all succumbed to the extreme conditions within an hour of landing, and only a few direct pictures of the surface have been taken. But radar maps have been made from orbiting spacecraft, showing numerous volcanoes and other related features.

Views of Venus

Venus is not as rewarding to observe as the Moon, Mars or Jupiter. Vague markings are visible where the clouds are slightly thinned, but they are not thin enough to allow views of the surface. Venus observers use a wide variety of color filters, probably a wider range than used for any other planet, to try to make the markings more prominent. The dark markings of Venus are typically shaped like a sideways "Y" and take about four Earth days to rotate around the planet.

As the planet is so dazzling, many people prefer to observe it during the day, or at least in twilight, as Venus is then higher in the sky and the contrast between it and the sky is not so great. Venus is easy enough to find by daylight if you know exactly where to point the telescope. If it is already polar-aligned, you can use the setting circles with which most equatorial mounts are equipped. It is best to leave the telescope set up from the previous night, or at least mark the positions of the tripod legs on the ground so that you can put it back already aligned. You can then use the declination scale straight away, but will have to point the (capped) telescope at the Sun to set the RA circle. A computer mapping program will give you the correct coordinates for Venus and the Sun.

Users of Go To telescopes have a slight problem as there are no stars available for aligning the telescope. However, if you take great care over setting up the telescope's "home position" – such as pointing it north and level to begin with – you should

▲ *Venus is an unmistakable sight in the twilight sky, appearing before any other stars or planets are visible.*

▶ *Venus seen in close-up by the ultraviolet camera aboard the US Pioneer spacecraft, showing the cloud patterns very clearly.*

be able to get an approximate alignment which will allow you to see Venus in the finder. Simply press "Enter" when the instrument points to each alignment star, even though you cannot see the star to center it.

As well as the markings, there are some other features that Venus observers like to look for. The terminator line – the edge of the sunlit part – is usually smooth, but on occasions it may have kinks. There may also be bright areas, known as cusp caps, around the cusps or tips of the crescent. One peculiarity is that when Venus should be at exactly half phase or *dichotomy*, with the terminator as a perfect straight line, it usually shows a lesser phase, as a slight crescent. Another interesting phenomenon is what is called the *ashen light*, in which the unilluminated side of Venus glows faintly when the planet is a thin crescent.

Mars

Other planets may be bigger and more spectacular, but everyone wants to see Mars through a telescope. Truth to tell, it can be rather a disappointment, but a glimpse of those mysterious dark markings, once thought to be vegetation, and the glistening polar caps, so reminiscent of Earth's own, is enough to fire the imagination. No matter that all you can see is a tiny disk jumping about in the air currents over your telescope, this is Mars, possible abode of life.

Mars is often called the Red Planet, but in reality the planet is more of a pale orange color, both to the naked eye and through a telescope. Its color is most obvious at opposition, when it can become even brighter than Jupiter, and at all times it is noticeably reddish in color. Though traditionally the planet is associated with the color of blood, which has led to its links with war, you should not expect to see a strongly colored object. Binoculars show the color well, but they are not powerful enough to show anything of the disk.

Mars is a rather small planet, and it is often a long way from Earth. Because Earth and Mars are engaged in a continual race around the Sun, we only see it at all well every two years when there is a close approach and it is at opposition (see page 12). Only at these times can you hope to see much detail on the planet. At a close opposition, Mars can be as large as 25 arc seconds across, compared with 47 arc seconds for a typical opposition of Jupiter. For much of the time when it is on the far side of its orbit from Earth, however, it can be less than 4 arc seconds across. In such circumstances very little can be seen, particularly with small telescopes. So you have to choose your time carefully and be patient if you want to see much detail on Mars.

▲ *The reddish color of Mars is obvious when compared with a blue-white star, such as Spica (the lower object).*

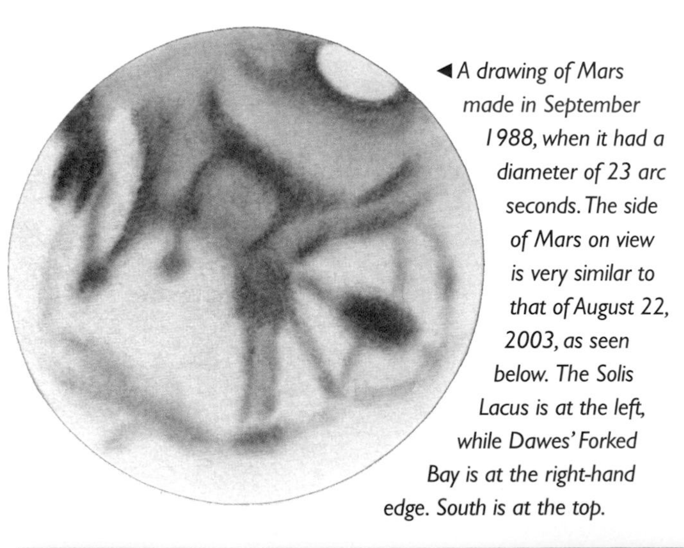

◀ *A drawing of Mars made in September 1988, when it had a diameter of 23 arc seconds. The side of Mars on view is very similar to that of August 22, 2003, as seen below. The Solis Lacus is at the left, while Dawes' Forked Bay is at the right-hand edge. South is at the top.*

For months after its first appearance in the early morning sky Mars remains a morning object. Then for a month or two on either side of opposition it is a worthwhile target to observe. Thereafter it starts to dwindle in size, but an odd thing happens: it seems reluctant to leave the evening sky, and remains visible for many months, though tantalizingly small. In actual fact Mars lingers roughly as long in the evening sky as it does in the morning, but it is just more obvious to most people. As the planet moves farther from Earth its phase changes, and at times only about 85 percent of the globe is illuminated.

▼ *The changes in size, distance, phase, and visible detail of Mars over five months from 2003 to 2004. It remained visible in the evening sky for a further eight months, eventually being half the size shown in January, before becoming lost in the twilight.*

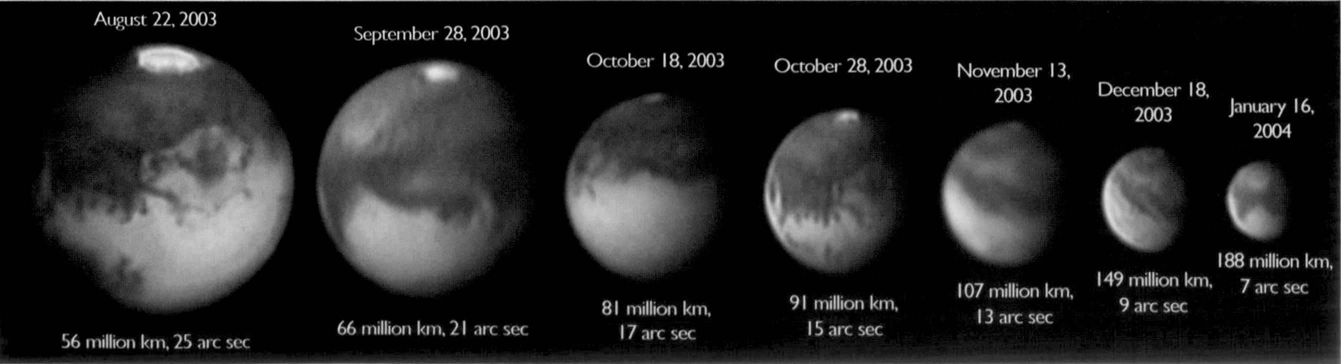

Mars through a telescope

Photos of Mars often have their contrast enhanced to bring out the detail, and this tends to raise people's expectations of what they will see. The contrast of the features is quite low, and their outlines ill-defined, so often all you can see on Mars, even when it is close, is a slightly darker tint on part of the disk. You may even wonder if your eyes are simply playing tricks, but if you can manage a magnification of at least 70 or so, you should be able to distinguish some features in good seeing. Mars, being small, is particularly sensitive to bad seeing, and sometimes you may even see several overlapping images. If you allow your telescope and the conditions to settle down, you may start to see more detail. Higher powers – the highest you can manage – will help if your telescope and the seeing are good enough, but Mars is a challenge for any small telescope.

The markings on Mars are more or less permanent, so you can compare what you see through the eyepiece with a map in order to establish which part of the planet you are viewing. But one confusing factor is the varying tilt of Mars' axis as seen from Earth. Like Earth, the axis of Mars is tilted to the ecliptic, and by about the same amount, 25°. This means that the same feature can appear at a different position on the planet's globe from apparition to apparition, depending on whether we are seeing predominantly the northern or the southern hemisphere. This makes a surprising difference to the recognizability of features. A good way round this is to use a computer-generated map of the globe, either using planetarium-type software (see page 126) or online. Such a map will show you the current aspect of the globe, though there is a risk that once you know what you should be seeing, you will let that image influence your observation.

As a day on Mars is just 37 minutes longer than our own, changes in the appearance of the planet due to its rotation are visible within a fairly short space of time. What's more, if you observe at the same time on successive nights you will see virtually the same side of the planet on each occasion. Only by observing at a considerably different time, or by observing a week or two later, will you see a different side of the planet.

To see a globe showing the features visible on the surface of Mars at any chosen time, go to www.calsky.com, and click on Planets: Mars: Apparent View/Data.

Although the major features of Mars remain from year to year, changes in the fine details are taking place all the time as a result of dust storms on the planet. Some features have changed considerably over the years, so there is no certainty about the exact appearance. Mars has a thin carbon dioxide atmosphere and dust storms are a common feature. They usually cover just a small part of the surface, so a particular feature may appear less distinct than it should, or may even be completely absent. Dust storms tend to be more common around the Martian perihelion, which is when it is closest to the Sun. This may not coincide with Mars' closest approach to Earth, however.

The orbit of Mars is noticeably eccentric – that it to say, the ellipse of its orbit is more flattened than that of the Earth's orbit – so its distance from the Sun varies quite considerably over its year. As a result the solar heating of the surface also varies, and is at its maximum at perihelion, which occurs when it is midsummer in Mars' southern hemisphere. Occasionally, dust storms may even develop to cover the entire planet with a yellowish haze. In 1971, no surface detail was visible on the planet for several weeks.

The polar caps, too, change with the seasons. They are easily visible with even a small telescope when the planet is close, though when the seeing is bad they can appear as little more than a brightening at one edge of the disk. The northern cap is less changeable than the southern one, and the southern one can almost disappear in its summer. As the

25.1″ 13.7″

▲ Mars, photographed at the very close opposition of 2003 (left) with a 250 mm telescope, and the distant opposition of 2012 through a 355 mm telescope, with diameters shown in arc seconds. The changed inclination of the axis makes a big difference to the features on view. South is at the top. The planet was lower in the sky in 2003 than in 2012, and was photographed with a smaller telescope, hence the difference in image quality.

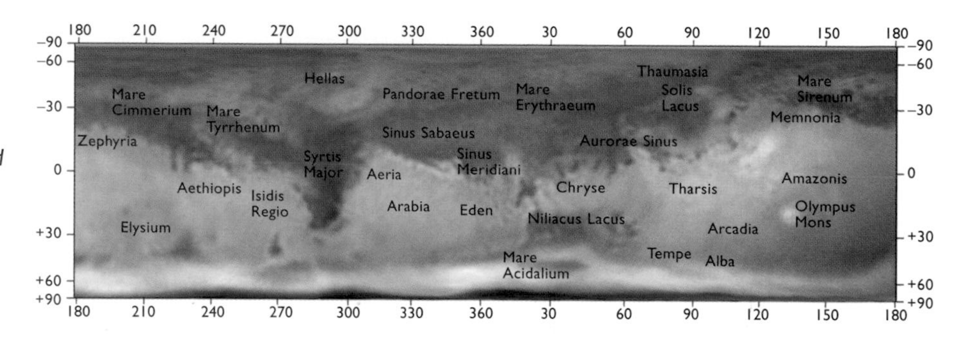

▼ A map of Mars from images by Damian Peach using a 355 mm telescope. This is known as an albedo map because it shows brightness variations only, unlike those made from spacecraft images which often show surface features. South is at the top.

caps shrink they develop ragged edges, which appear through the telescope as tongues or even islands of brightness. Usually only one cap is visible at a time, as a result of the planet's tilt.

Though debates rage over just how much water there is on Mars and where most of it lies, it definitely does exist on the planet. Clouds of water or carbon dioxide ice can form. They are distinguishable from dust storms by their white or bluish color and can occasionally be seen at the eyepiece.

In addition to the named features on Mars, there are one or two popular names that are occasionally used but which do not appear on maps. Dawes' Bay or the Forked Bay is one, being the distinctive double dark area of Meridiani Sinus. This is the zero longitude point on Mars' grid system, hence its official name. Another is Solis Lacus, also known as the "Eye of Mars," which in some years looks like an eye with eyebrow when seen in a telescope that inverts the view. The famous Valles Marineris canyon marks the lower lid of the eye, though this is not visible through a telescope. The view of Mars that we see at the eyepiece is essentially an *albedo* view – that is, the view is of brightness variations only and not of the actual relief. (The term *albedo* refers to the reflectivity of a surface.) So the craters and other features are not properly visible. However, a feature once named Nix Olympica, the Olympian Snows, is now known to be a huge volcano, and has been renamed Olympus Mons or Mount Olympus. You may find both terms used.

In the late 19th and early 20th century, many people believed that they could see numerous straight markings on Mars, which were termed canals. These were mostly wishful thinking, and resulted from a tendency for the eye to join unrelated blobs. There are some linear features, but they have nothing to do with intelligent life, as was once believed.

▲ *The western edge of the 22 km (14-mile) Endeavour crater on Mars, photographed by the Opportunity rover in August 2011. This long-lived rover landed on the planet in 2004, and has sent back thousands of panoramas and close-ups of the surface.*

As a planet, Mars is now dry and dusty but there are many signs that water once flowed in abundance. The wet periods were probably brief, but there is a chance that primitive life may have gained a toehold at some stage and remains there, probably below the surface.

Jupiter

Jupiter is a favorite target for amateur astronomers. Although it is much farther away than the inner planets, Mercury, Venus and Mars, it is so large that it always presents a sizable disk when viewed through a telescope. In fact, you can probably see that it has a disk using binoculars, though you are unlikely to be able to see any detail.

Once you know where Jupiter is in the sky, you can keep track of its movement very easily. It takes almost 12 years to orbit the Sun once, so its movement through the sky month by month is small compared with that of the inner planets. So next year it will be in the adjacent constellation to where you see it this year, and it will be visible at roughly the same time of year. Like the other planets Jupiter marches from west to east over the months, though the Earth's daily rotation makes it move from east to west as its part of the sky as a whole rises and sets. Jupiter can also go through retrograde loops (see page 12), though it remains in the same part of the sky while doing so.

▲ *Jupiter appears brighter than any star. It was in the constellation of Capricornus when this photo was taken.*

Views of Jupiter

To the naked eye, Jupiter is white and brighter than any star. Venus also meets this description, but is even brighter. If it is high in the sky late at night, it can only be Jupiter.

▲ *In a small telescope, Jupiter's satellites and belts are easily visible.*

A glance through binoculars immediately confirms that it is Jupiter, because it is accompanied by its four bright moons, which are easily spotted even using binocular magnifications. Not all of the moons are always visible but only very rarely can none be seen at all using binoculars.

With a telescope and a magnification of only about 20 or more, detail starts to appear. Because Jupiter is a gas planet, there is no solid surface, and all we see is the top of its opaque atmosphere. However, there are general features that remain constant over the years. To start with you should see what appear to be two darkish stripes across the planet, parallel to the equator. You will also notice that the planet is slightly flattened from a circle. This is a sign of its rapid rotation – this giant planet spins in just under 10 hours.

The dark stripes are referred to as belts, and the most prominent are usually those on either side of the equator, but there are other minor belts as well. Sometimes the South Equatorial Belt fades, and Jupiter appears to have only one equatorial belt. As you gaze at the planet you should be able to make out more detail. The pictures of Jupiter in books often have their contrast and color enhanced to make the belts appear more distinct, so the planet may be a little disappointing the first time you see it. Novice viewers may not even see the belts at all, but as with all astronomical observing you have to train yourself to study the object carefully if you are to get the most out of it.

Belts and zones

To see the belts and zones in more detail try increasing the magnification, though not so much that the planet becomes dim. Even a good 60 mm telescope will start to show you a bit more detail as you do so, and with a magnification of about 45 the planet will appear as large as the Moon does in the sky to the naked eye. With a 75 mm telescope more details become visible, and you should be able to see that the belts are not just plain stripes but have irregularities in them. A telescope of 100 mm or more starts to show spots, both light and dark. A magnification of 90 will give you a good general view of the planet, but higher powers are helpful if you want to see the fine detail.

The lighter areas are known as zones, and they and the belts have a nomenclature of their own, as shown below. There are also long-lived features that have their own names, yet which don't appear on general diagrams. Jupiter's atmosphere is one great raging storm, but there are individual disturbances which have remained for many years. The most famous storm is the Great Red Spot in the South Tropical Zone, a reddish lozenge that has been there since at least the 1820s and probably back to the 17th century. Its color and size vary over the years – in the 19th century it was about a third larger than it is at present, and was a darker color. At the beginning of the 21st century it is hardly any redder than the belts. It sits in a lighter area known as the Red Spot Hollow. Other smaller spots may persist for several years at a time and receive their own designations.

Drawing the planet

The planet's rapid rotation becomes very obvious when you observe it for more than a few minutes at a time, particularly if you are trying to make a drawing. In a short time, the features that were in the center of the disk start to move to one side, and new features appear at the opposite edge. If you are trying to draw the planet, begin with the leading edge, containing the features that will disappear first. Observers learn to sketch the planet in outline quickly rather than laboring over subtle shadings. It is best to start with an oval outline or blank, which these days is most easily prepared by scanning an existing drawing into the computer and deleting the details before printing it out.

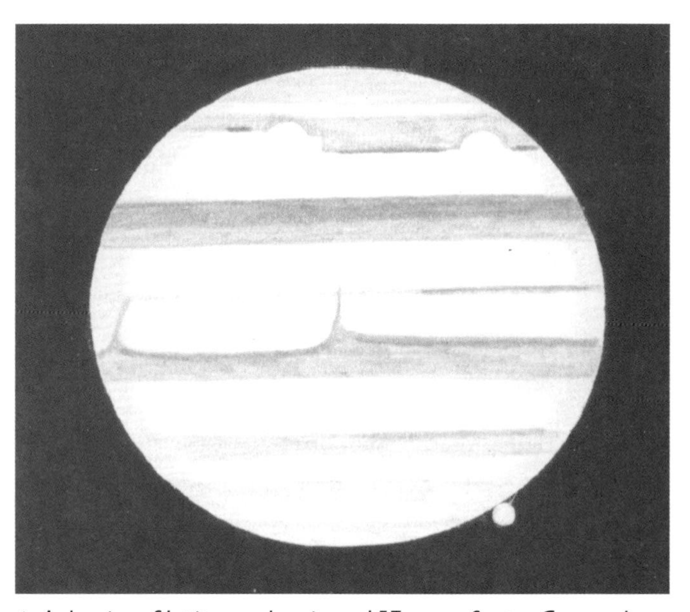

▲ *A drawing of Jupiter made using a 157 mm refractor. Ganymede has just emerged from behind the disk.*

Sometimes you may see what looks like a particularly dark, circular spot on the planet. This is almost certainly the shadow of one of the major satellites, which will be nearby, though if it is in front of Jupiter's disk rather than to one side you may not be able to pick it out.

Jupiter close-up

Jupiter and its satellites have been studied in detail by several spacecraft, most recently Galileo, which orbited the planet between 1997 and 2003 and dropped a probe into the atmosphere. The darker belts are believed to be lower in altitude than the bright zones, and there are probably three layers of clouds in all. The lowest consists of water ice, then above that is a layer of ammonia and hydrogen sulfide. The uppermost clouds are ammonia ice. But a lot remains to be discovered about the chemistry of Jupiter's clouds.

◀ *The nomenclature of Jupiter's belts and zones. The photograph has south at the top, as seen in a telescope, and includes the Great Red Spot.*

◄ A composite of Jupiter and its Galilean satellites, based on photographs from the Galileo orbiter.

The four large or Galilean satellites, so-called because they were first seen by Galileo Galilei in 1610, also hold their share of secrets. The innermost, Io, is so strained by Jupiter's enormous gravity that its interior is molten, and sulfur volcanoes constantly erupt on its surface. The next out, icy Europa, appears to have a frozen surface but possibly has a liquid ocean beneath. There is speculation that it may even harbor primitive life forms. Farther out, Ganymede is the largest satellite in the Solar System and is bigger even than the planet Mercury, while Callisto is only slightly smaller. Both have cratered icy surfaces.

Saturn

Saturn is an amazing sight through a telescope. If we had never seen a planet with a ring round it, we would not believe such a thing was possible. Many small telescopes are sold on the basis that they will show the rings of Saturn, but in fact all but the worst telescope should be able to accomplish this feat. A magnification of 20 is all you need to glimpse the rings, and even binoculars will reveal that the planet is not a simple disk but appears elongated.

Being almost twice as far from the Sun as Jupiter, it takes nearly 30 years to orbit the Sun, so it moves only slowly from constellation to constellation and remains visible at roughly the same time of year for years at a time. Its plod around the heavens gives rise to the word *saturnine*, and in ancient times it was the slowest known planet.

Views of Saturn

To the naked eye it appears as a bright, noticeably yellowish star. There are a few stars brighter than Saturn, but being a planet it rarely twinkles, so it is easy to distinguish. Though its disk is about half the size of that of Jupiter, the rings bring it up to just about the same diameter as Jupiter so even a small telescope will show it well. But if it were not for the rings, Saturn would be a great disappointment. Its disk is bland compared with Jupiter's, with just a few belts and zones, and comparatively few other features. Saturn appears the same at first glance whenever you see it, but there is enough going on to interest its devotees.

▲ Saturn is the bright object above center, here seen within the non-zodiacal constellation of Cetus. The bright star below it nearer the horizon is Fomalhaut, in Piscis Austrinus.

▲ Changes in the tilt of Saturn's rings, as recorded over 15 years.

One thing that clearly changes is the inclination of the planet to the Earth. While Jupiter is virtually always seen edge-on, because its polar axis is inclined at only 3° to the vertical of the ecliptic, Saturn's axis is inclined at nearly 27°. This means that as it progresses round the Sun we see it inclined at different angles. The cycle from edge-on to fully inclined either north or south takes about $7\frac{1}{2}$ years, then it returns to edge-on and starts to incline at the opposite angle and show us its other pole. The rings are the most obvious sign of this change, and each year they appear noticeably different.

Of Saturn's many moons, six or seven are easily visible in amateur telescopes. They orbit in the plane of the rings, so usually they are spread out somewhat, rather than being in a line like Jupiter's Galilean moons. Only the brightest, Titan, is visible in the smallest telescopes, and you may even glimpse it in binoculars.

Observing Saturn

Many people are happy just to gaze at Saturn, for it is a spectacular sight at all times. But if you want to take your interest a step further, you could make a drawing. This is not as easy as it sounds, because whereas in the case of the other planets a simple blank disk is adequate, in the case of Saturn the changing aspect of the rings is a challenge. Observing sections of societies such as the British Astronomical Association (BAA) make blanks available to their members, and they can be obtained online for home printing for any chosen date from the Association of Lunar and Planetary Observers at http://www.alpo-astronomy.org/saturn/satfrms.html.

Although the angle of the rings varies only slightly throughout the year, you will see a considerable difference in their appearance during the apparition as a result of the changing illumination. Around the time of opposition there are no shadows and the globe is at full phase, but on either side of opposition you may see the shadow of the globe on the far side of the rings. It is also possible to see the shadow of the near side of the rings on the globe.

The rings are actually composed of millions of tiny particles of ice and rock, all orbiting in a very thin plane, no more than a kilometer thick. As a result, when the rings are seen edge-on, they virtually disappear.

Although the globe itself is bland compared with that of Jupiter, it is by no means unchanging. There are variations in the color and intensity of the various belts and zones, and from time to time white spots appear, particularly every 30 years or so when the planet's north pole is tilted toward the Sun.

In recent years, amateur astronomers have been able to observe Saturn using CCDs and webcams. Image processing can enhance features of low contrast, and images taken from back gardens with moderate aperture telescopes can reveal features that might previously have been at the borderline of visibility.

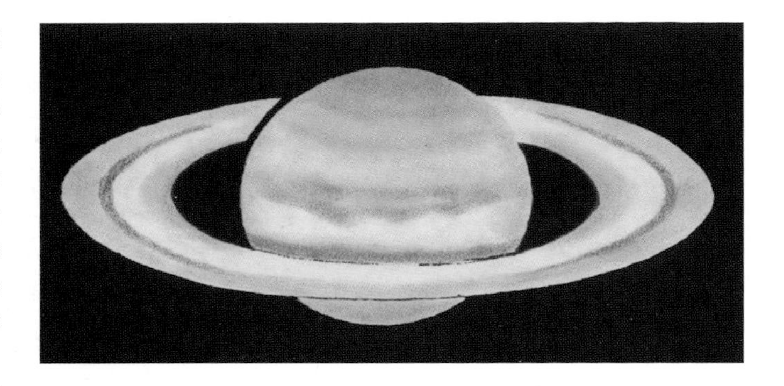

▲ *An amateur drawing of Saturn made in August 1991 with a 157 mm refractor and magnifications of 152 and 213. It shows a considerable amount of detail in the North Equatorial Belt.*

▲ *A rare storm on Saturn, photographed by the Cassini spacecraft in February 2011. This was the largest event seen on the planet since 1933, though even from Cassini the planet does not normally show much detail.*

◀ *The same storm, photographed with a 355 mm telescope from Earth a month later. Some Saturnian storms have been spotted by amateurs before they were seen by Cassini, which spends much time imaging the satellites. This photograph was taken just a few days before opposition, when the rings appear much brighter than usual because the icy particles reflect sunlight directly back in the same way as reflective road signs.*

Uranus and Neptune

These two outer planets are rarely observed though they are both comparatively easy to locate. Uranus is sometimes just visible to the naked eye as a very faint star, but it was overlooked completely until 1781, when an amateur astronomer named William Herschel using a homemade telescope noticed that it has a disk rather than being starlike. It was the first planet ever to be discovered, rather than being known since from ancient times.

Neptune was located after it was found that Uranus was deviating from the orbit that had been calculated, and it became obvious that a more distant planet was pulling on it. The planet, Neptune, was eventually tracked down in 1846. At about 8th magnitude, it is well within the visibility of binoculars and small telescopes, but it appears starlike except in larger telescopes.

Both planets are similar in composition, and while they are often referred to as gas giants, along the lines of Jupiter and Saturn, they have denser interiors, with a large proportion of water rather than hydrogen gas.

Viewing Uranus and Neptune

The best way to locate these planets is either to consult a computer sky mapping program in order to print out a map to use at the telescope, or to use a Go To telescope. Either way, you are looking for a starlike object that appears greener or bluer than the other objects. You will not see a disk at low magnifications. If you have a star map that shows stars fainter than the planet's brightness, you may be able to establish which one it is because it will be in addition to those shown on the map. Both planets move more slowly from night to night than the bright planets, but after only one night there is still a noticeable change compared with the background stars. Currently both planets are well away from the Milky Way, and will remain so for many years, so there are comparatively few stars in the field of view.

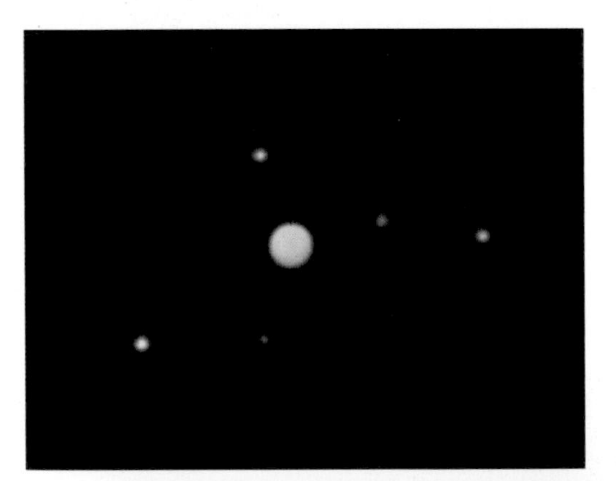

◀ *An amateur CCD photo of Uranus and its brighter satellites, taken with a 355 mm SCT. Though the planet's disk is visible with much smaller instruments, the satellites are 14th magnitude or fainter and can only be seen in large telescopes.*

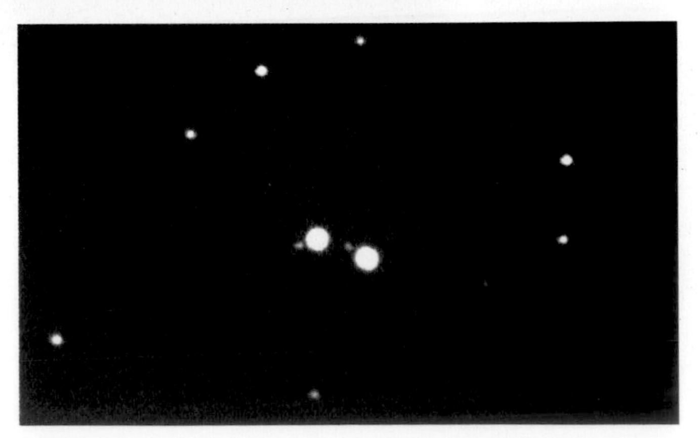

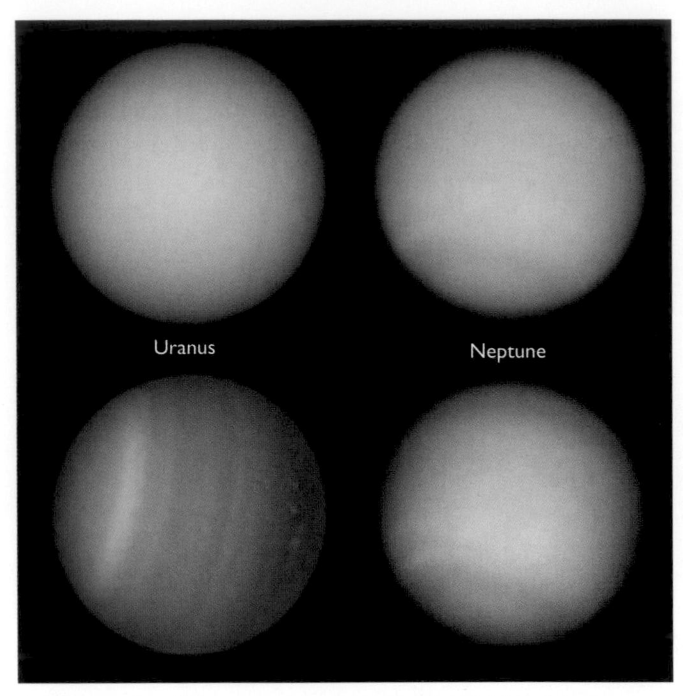

▲ *Photographs of Uranus and Neptune in natural (above) and enhanced color (below), taken in August 2003 by the Hubble Space Telescope. Even at this resolution very little detail is visible.*

In binoculars Uranus is easy to spot, and you can pick it out straight away by its distinctly bluish or greenish color. No star ever appears green – it is a physical impossibility, despite occasional claims to the contrary. You will need a moderate aperture (about 150 mm or larger is adequate) with a power of 50 or more and a night of fairly good seeing to view the disk of Uranus well. It is only 3 or 4 arc seconds across – less than a tenth that of Jupiter – and is much dimmer than that of any of the bright planets. But when you do spot this tiny blue disk you are reminded of the Hubble Space Telescope photos that show virtually the same thing, but larger. There is almost no detail to be seen on Uranus, even using the HST, so your own glimpse can be quite rewarding.

Neptune is also well within the reach of binoculars, but being fainter requires more of a search and less certainty that you have found it. Even with a Go To telescope you may not know which of several objects in the low-power field of view is Neptune, so again you may need to compare your view with a detailed map to be sure. You will need a telescope with a power of about 100 to show its bluish disk, of about 2 arc seconds in diameter, which like that of Uranus does not show any detail with amateur telescopes.

Probably every observer makes the effort to look at these planets very occasionally, then forgets about them for another few years. They are left in peace way out in the depths of the Solar System, largely unwatched by human eyes. By comparison, Saturn is probably being viewed by people round the clock. So spare a thought for these lonely planets and give them a friendly look from time to time. You may be surprised by how much you can see and by the strong color of their disks, which is caused by methane in their atmospheres.

◀ *Neptune and its largest satellite, Triton, photographed with a CCD on two separate nights. The planet (center) has moved between the two superimposed exposures, and Triton has also moved in its orbit around the planet. The planet's image is overexposed in order to show Triton, which is magnitude 13.5.*

Pluto

For years, Pluto was regarded as the ninth planet, but these days more objects of similar or even larger size (though not brighter) have been discovered at the fringes of the Solar System. In 2006 it was reclassified as a dwarf planet, along with the largest asteroid, Ceres. But for many people, Pluto is still a planet and they want to find it if their telescope is large enough.

As it is 14th magnitude, you need at least a 250 mm telescope just to glimpse Pluto as a starlike object at the limit of the telescope's range. Unlike the major planets, it is far too small to show a disk – so at a glance it is not obvious which of many starlike objects of similar brightness is Pluto. This is made more difficult by Pluto's current position against the background of the southern Milky Way, where it will remain for many years. A computer-plotted star map going down to at least 16th magnitude is the best way of finding it, as even an accurately aligned Go To telescope will not show which of several faint stars in the area is Pluto. With a map, you may be able to spot which one is the interloper but even then you will need a second night's observation to be sure.

► Pluto photographed using a CCD with a 300 mm telescope. Images for two nights have been superimposed, so the planet's image has moved slightly.

Observing asteroids

There are literally tens of thousands of asteroids – rocky bodies that orbit mostly between Mars and Jupiter. After the Sun formed some 5 billion years ago, it is believed to have been surrounded by a disk of accumulated material with the densest bodies closer in and the lighter ones farther out. The dense bodies gathered to form the inner rocky planets, while farther out the lighter material formed the giant planets and their icy satellites. Beyond Mars there was not enough rocky material to form a single planet, and what was left over remains as the asteroid belt.

The brightest asteroids are easily visible with binoculars and small telescopes, and the fainter you can observe the more you can see. To find them, you can use the same general methods as for Pluto, either using a computer-plotted star map or a Go To telescope. As asteroids are comparatively close, you need wait only an hour or two before the asteroid reveals itself by its motion, even if it is not obvious from its position on the map.

Observing comets

Bright comets are few and far between. Comet Hale-Bopp in 1997 was the best known of recent years. Virtually everyone was able to see it, even from city locations, whether or not they were familiar with astronomy. But most comets are much less spectacular. Every year a dozen or so new ones are found, but only rarely are they visible without optical aid and even more rarely do they sport a noticeable tail. Most of the comets never reach the attention of the public, and details are only to be found on astronomical websites rather than in the national press.

Comets are actually fairly small icy bodies only a few kilometers in diameter. They only become easily visible if their orbits bring them into the inner Solar System, and though there are some comets that have orbits similar to those of asteroids, most have very eccentric elliptical orbits and spend most of their time in the depths of the outer Solar System where they originate. Only for a few weeks at a time are they close enough to the Sun to become observable, when the increased solar heating results in their ice turning to gas. At this point the cometary body or nucleus becomes surrounded by a halo known as the *coma*, with tails of gas or dust or both stretching away from the nucleus.

The nucleus itself is too small to be observed directly. The starlike point often seen within the coma is really the brightest part of the gas emission, and is known as the *pseudo-nucleus*.

All known comets are in orbit round the Sun, but those with short periods of a few years are rarely bright enough to be noteworthy. Only the unpredictable comets that are unknown until they suddenly arrive in the Sun's vicinity are likely to become spectacular, the only exception being Halley's comet, which has a period of about 76 years and is not due back until the middle of the 21st century.

► Comet Hale-Bopp in July 1996, when it was a binocular object. It shows the typical appearance of a comet – a hazy blob, with no tail.

▲ By March 1997, Comet Hale-Bopp was a spectacular sight in the night sky and was visible even from city centers. The white tail is dust and the blue tail is gas.

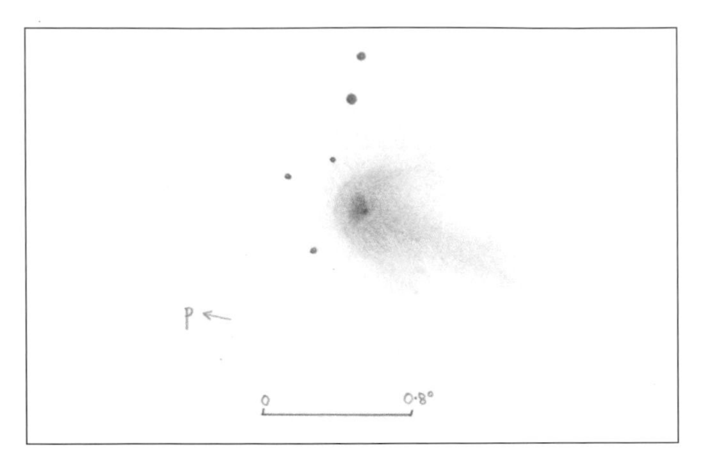

▲ *An amateur drawing of Comet NEAT (C/2001 Q4). It appeared in 2004 at 3rd magnitude in the twilight evening sky. Drawings are often able to show a wider brightness range than photographs, and here the starlike pseudo-nucleus is visible within the coma.*

When a bright comet appears it can best be observed with ordinary binoculars. But the more run-of-the-mill faint comets require large binoculars or telescopes. Astronomy magazines and websites give either maps of their path, predictions of their position from day to day, or just the details of the orbit (known as *orbital elements*) which you can insert into computer sky mapping software to produce your own predictions. A set of predictions is known as an *ephemeris*. For more details see the website http://www.minorplanetcenter.org/iau/Ephemerides/Comets/.

There are six orbital elements that precisely define the orbit of a comet or indeed any other body, such as an asteroid. The time and date of perihelion (the closest passage to the Sun) are known as T, which defines the position of the body in time. The size and shape of the orbit are given by q, the semimajor axis, and e, the eccentricity of the orbit. Finally, the body's direction in space is given by three parameters called the longitude of perihelion, ω, the longitude of the ascending node, Ω, and the orbit's inclination to the ecliptic, i. In addition, the magnitude of the body may be given by two figures, H and G. These figures are given for all comets by the International Astronomical Union's Central Bureau for Astronomical Telegrams, which is the world clearing-house for discoveries.

Eclipses

Eclipses of the Sun and Moon are usually well covered in the press these days as they can be seen without specialist equipment. But as always there are misconceptions about what will happen, and in the case of solar eclipses the information is sometimes downright misleading.

It is a great coincidence that the Sun and Moon happen to be about the same size in the sky, around 30 arc minutes. Solar eclipses occur when the Moon goes directly across the face of the Sun, while eclipses of the Moon take place when the Moon passes through Earth's shadow.

As the Moon goes round the Earth once a month, you might expect there to be an eclipse of some sort every couple of weeks, but of course this is not the case. The reason is that the tracks of the Sun and Moon through the sky (or more correctly, the orbits of the Earth around the Sun, and the Moon around the Earth) are tilted with respect to each other. So only when the Sun and Moon happen to be close to the place where the two tracks cross can there be an eclipse. The tracks cross every 173 days, or just under six months, so for a week or two on either side of the date when the Sun and Moon are close to their crossing points there is a chance of an eclipse of either the Sun or Moon. Whether the event will be visible from your location is another matter, as you must be within the Moon's shadow on the Earth (about 3500 km across) to witness any sort of a solar eclipse. To see a lunar eclipse, however, it simply needs to be night-time at your location. So from any location, lunar eclipses are far more common than solar ones.

A moment's thought will tell you that the Moon must be new for there to be a solar eclipse, and full for there to be a lunar one (see page 43). During a lunar eclipse the Sun and Moon are exactly opposite each other in the sky.

Observing solar eclipses

Solar eclipses are seen as total by observers exactly within the shadow of the Moon, but partial by those near the edges. During the partial phases, the Sun appears to have a bite out of it, as the silhouette of the Moon begins to cover the brilliant solar disc. At this stage, you can only observe the Sun by using the same methods as at any other time – by projection or by using the correct dense solar filters, either over the aperture of the telescope or with the naked eye. At one time, people tended to use any material that reduced the visible brightness of the Sun, such as smoked glass, but we now know that such materials may allow infrared light to be transmitted. This can damage the eye just as readily as the white component, but as the Sun looks visually dim there is no reflex action to make you look away. Color filters come into the same category, along with black plastic bags. Only use approved solar filters to look at the Sun with the naked eye.

► *The total eclipse of August 11, 1999, as seen from Cornwall, UK. Prominences surround the Sun. The corona can be seen to a considerable distance from the edge of the Sun.*

▲ *Stages of the lunar eclipse of November 9, 2003. The Earth's shadow encroaches from the west as seen in the sky. At totality the Moon appears reddish, usually with a brighter segment if the Moon does not pass centrally through the Earth's shadow.*

When the Sun is totally eclipsed, however, its bright surface is hidden by the Moon and it is safe to view the spectacle, but only for the duration of the total stage itself. This can last anything from a few seconds to seven minutes or so, depending on the relative diameters of the Sun and Moon at the time. The outer layers of the Sun (or corona), which are normally too faint to be seen, become visible, as do any prominences that happen to be at the limb of the Sun. A few stars and the brighter planets appear in the sky. When the Moon either moves on to or off the solar disk, there may be a brief glimpse of the photosphere through a deep lunar valley. This is known as the "diamond ring" effect.

Observing lunar eclipses

Fortunately there is no direct danger associated with observing lunar eclipses. During the event, the Earth's shadow slowly encroaches on the Moon and then moves away again. If the Moon passes through only the outer edge of the Earth's shadow, the *penumbra*, it may just appear slightly dimmer than usual at one edge. It may partially pass through the black part of the shadow, in which case we see a *partial eclipse*, or entirely into the shadow, which gives a *total eclipse*.

During totality, the Moon often turns a dark shade of red, orange or brown. This is because sunlight is bent and absorbed by the edge of the Earth's atmosphere, with the red component being most likely to reach the Moon. The darkness or lightness of a total lunar eclipse is always an unknown quantity beforehand and adds to the interest and spectacle of the event. Through a telescope, the lunar features usually remain dimly visible.

Occultations

From time to time, the Moon passes directly in front of a star, temporarily hiding it from view. This is known as an *occultation*, and precise timings of such events are of some use in keeping accurate track of the Moon's movement. Of course, the Moon is passing in front of stars all the time, but stars fainter than about magnitude 8 are virtually unobservable visually when so close to the bright Moon.

When a star is occulted, it disappears instantly because the Moon has no atmosphere to dim the star's light before it is occulted. This is why timings can be so precise. It is easiest to observe occultations by the dark limb of the Moon, as you can watch the star until it disappears and there is no glare to worry about. But reappearances or occultations by the bright limb are subject to greater errors because you do not know precisely where the star will reappear, and the bright limb creates glare, which may mask the instant of reappearance. Lists of observable occultations from any location worldwide are available from the International Occultation Timing Association website at www.lunar-occultations.com/iota/iotandx.htm. Occasionally, the planets may be occulted by the Moon, and occultations of stars by planets or even asteroids are also possible.

Meteors

Watching for meteors – popularly known as shooting stars – is the astronomical equivalent of fishing. You never know what will happen next, and there can be long periods when nothing happens. But there is always the chance that in the very next instant you will get the big one that makes it all worthwhile.

Meteors can occur at any time. They are tiny grains of interplanetary dust that burn up in a sudden trail of light when they collide with the Earth's atmosphere at a height of about 80 to 100 km. The bodies that cause meteors are called meteoroids, and they are generally very small, low density objects, more like grains of instant coffee than grains of sand.

Meteors mostly come from the tails of comets, so the shooting star that you see dashing through the sky is the destruction of a tiny piece of material that probably originated billions of years ago in a distant star. It then became part of the formation of the Solar System and found itself in the body of an icy comet. Eventually it was released from the comet on a passage close to the Sun and has been orbiting for millions of years until its vaporization before your eyes.

The orbits of comets are strewn with the dust released over a long period of time. This dust spreads outward from the path of the comet as a result of various forces, such as the pressure from particles emitted by the Sun and the absorption and re-emission of sunlight. Eventually the dust becomes randomly spread throughout the Solar System, and it is this dust that gives rise to the *sporadic* meteors – that is, meteors that occur at random.

If the Earth passes close to the spreading cloud of dust from a comet, however, there will be a greatly increased rate of collisions and we see a *meteor shower*. In a shower, meteors appear to come from a particular point in the sky because of the effects of perspective. This point is known as the *radiant*, and the shower of meteors is named after the constellation in which it lies. One of the most famous and regular showers, for example, is the Perseids. Each year, in the second week of August, increased numbers of meteors appear to radiate away from a point in the constellation of Perseus. The shower can be seen from anywhere that Perseus is above the horizon, with the greatest numbers visible when it is high in the sky. A list of the major meteor showers is given here.

METEOR SHOWERS

Shower	Maximum	Normal limits	Rate at maximum	Radiant RA	Radiant Dec	Remarks
Quadrantids	Jan 04	Jan 01–06	100?	$15^h 28^m$	+50°	Blue meteors with trains
Lyrids	Apr 22	Apr 19–25	10–15	$18^h 08^m$	+32°	Bright meteors
η-Aquarids	May 05	Apr 24–May 20	40	$22^h 20^m$	−01°	Broad maximum and multiple radiants
α-Scorpids	Apr 28 May 12	Apr 20–May 19	10	$16^h 32^m$ $16^h 04^m$	−24° −24°	Multiple radiants – long activity April to July
δ-Aquarids	July 28 Aug 06	July 15–Aug 20	20	$22^h 36^m$ $22^h 04^m$	−17° +02°	Double radiant – southern component is richer
Perseids	Aug 12	July 23–Aug 20	75	$03^h 04^m$	+58°	Rich shower, bright meteors with trains
Orionids	Oct 21	Oct 16–30	25	$06^h 24^m$	+15°	Fast meteors, many with trains; associated with Comet Halley
Taurids	Nov 03	Oct 20–Nov 30	10	$03^h 44^m$	+14°	Slow meteors, some fireballs
Leonids	Nov 17	Nov 15–20	20	$10^h 08^m$	+22°	Trains, rates declining after major storms in 1999–2002
Puppids-Velids	Dec 08 Dec 25	Late Nov to Jan	15	$09^h 00^m$ $09^h 20^m$	−48° −45°	Two of several radiants
Geminids	Dec 13	Dec 07–15	120	$07^h 28^m$	+32°	Many fireballs
Ursids	Dec 22	Dec 17–25	10	$14^h 28^m$	+78°	Increased activity in certain years

The number of meteors visible is given by the Zenithal Hourly Rate (ZHR), and this is the figure usually quoted in the media when talking about what you might see. For the Perseids, it is around 60–100 meteors an hour. But in practice you will almost never see as many, because as its name implies the ZHR is the number that you would see if the radiant were in the zenith (directly overhead), in perfect conditions, and for a single observer looking attentively at as much of the sky as possible. This never applies to the Perseids, because the radiant is never overhead during night hours, so the numbers seen will be reduced. Skies are rarely perfect, and it is easy to miss the fainter meteors, which are often more plentiful than the bright ones. Most people would be happy to see a half or even a quarter of the stated number each hour, though a group of people observing in good conditions could see more meteors an hour than the ZHR figure would suggest.

Meteors are more common during the second half of the year than the first half, for reasons that are not fully known. Also, meteor rates increase toward dawn, because at that time the side of the Earth you are on is heading directly into the path of the meteors. Patience is often necessary if you want to see a shooting star, even on the night of a good shower.

In addition to meteors, occasionally much brighter objects dash through the sky. Anything brighter than magnitude −5 at its brightest is termed a *fireball*, and if it explodes it may be called a *bolide*. Fireballs may occur during a regular meteor shower as particularly bright ordinary meteors, but those that appear outside normal shower times, or do not come from the expected radiant, may originate within the asteroid belt rather than from a comet. Instead of being composed of rather crumbly dust grains, fireballs are likely to be chunks of rock or even metal. If they survive their descent, they are called *meteorites*. Thousands of meteorites fall to Earth daily, but only a tiny fraction of that number is ever recovered because most fall in the sea or over unpopulated territory. When a very bright fireball appears, people are asked to describe as accurately as possible its path through the sky in the hope that its possible drop point can be found and a meteorite recovered. Very often, however, people believe that a run-of-the-mill bright meteor or fireball is lower than it actually is, and associate a nearby chance noise with its impact.

▲ *A Geminid meteor below Orion. Photographing meteors is a matter of leaving the shutter open and hoping that one will appear. Only the very brightest will record successfully.*

Aurorae

Although aurorae – also called the Northern or Southern Lights – are a feature of the polar regions, from time to time they can be seen from lower latitudes. They are visible as colored glows in the sky, often with streamers or rays, and are the result of interactions between streams of particles from the Sun and Earth's magnetic field. They are most common at times of enhanced solar activity, such as sunspot maximum (see page 56), but are not out of the question at other times. The appearance of a large sunspot may herald an aurora, but there is no certainty, and sometimes aurorae occur when there is no obvious major solar activity.

Auroral activity is most intense in a band surrounding the Earth's geomagnetic poles, which are considerably offset from the geographic poles. The geomagnetic north pole is in Canada, and the geomagnetic south pole is near Vostok in Antarctica. So observers in Alaska, Canada or Tasmania are more likely to see the aurora than those in Europe or South America at similar latitudes. In Europe, people living in Scotland or Norway get far more opportunities to see the aurora than those in London or Paris, while in Rome only the most dramatic events are visible.

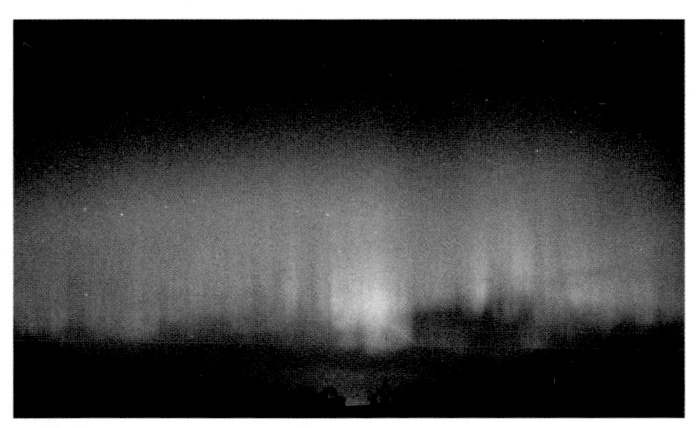

▲ *An aurora seen from southern England in April 2000. It shows the characteristic curtain appearance that is more typical of those seen nearer the poles. The color changes from green to purple with increasing height.*

An aurora may consist of little more than a pale glow along the poleward horizon, but in the more dramatic events there are noticeable rays or streamers. During a major display you may see the classic curtain-like appearance of these streamers, with the lower ends ending in definite curves. Sometimes the rays can appear to come from a point high in the sky, an effect known as a *corona* (not to be confused with the solar corona visible during a total solar eclipse).

The glow of an aurora occurs at an altitude of more than 100 km in the Earth's upper atmosphere, in a region normally considered to be space. In particular, oxygen molecules glow either green or red, with the red tending to be at higher altitudes than the green. As a result, observers in midlatitudes tend to see red aurorae rather than green, and only during a major display does the green become more obvious. Other colors are caused by other atmospheric gases, such as nitrogen. The colors are visible to the naked eye when the display is bright, but faint glows are easy to confuse with light pollution. A time exposure of 30 seconds or so with a sensitive digital camera will often reveal whether an aurora is in progress, as it shows the color much more strongly. Another clue is that light pollution inevitably shines on the underside of clouds, whereas clouds show dark in silhouette against an aurora. The auroral streamers change in intensity and position over a period of seconds, so photographs are best restricted to exposure times of less than 30 seconds, if possible, to avoid smearing of the detail.

There is no good way of predicting well in advance when an aurora will take place, although aurora alerts based on the release of particles from the Sun are now sent out by email to anyone interested. There are also continuous online monitors of geomagnetic activity and these will tell you whether it is worth even looking out of the window for enhanced auroral activity. For a listing of websites worldwide that provide such information, go to www.stargazing.org.uk.

The website www.spaceweather.com issues alerts by email and provides links to solar activity sites.

Noctilucent clouds

At one time these beautiful high-altitude clouds were hardly ever mentioned in observing guides because they were very rare events. But they are on the increase, and are now visible from lower latitudes than ever before, though exactly why no one is sure. Noctilucent clouds appear around midsummer at latitudes higher than 50° north or south. Almost all the reports received come from the northern hemisphere because there is very little populated territory below 50°S, so here we deal only with the northern-hemisphere appearance.

At first glance, noctilucent clouds (NLC) look like cirrus clouds in the late twilight, but they are visible long after sunset and only toward the northern horizon, where the sky is still light from late May to early August. They are silvery or bluish in color, and any ordinary clouds are seen in silhouette against them. NLCs are water vapor clouds, and they may be the result of condensation on tiny particles of dust from volcanoes, meteors or aircraft vapor trails. They are about 80 to 85 km above the Earth's surface, where normal clouds do not form. Because they are rather faint, NLCs are not seen during the day.

▲ *Noctilucent cloud seen from a London, UK, golf course. The star is Capella, which is usually near the horizon when noctilucent clouds are seen from the northern hemisphere.*

CHAPTER 7
STARS AND DEEP SKY OBJECTS

About stars

Each star that we see in the sky is a sun in its own right. Our Sun is a star of average brightness, so we can picture the others as being similar self-luminous globes, each one brilliant and a great source of heat and light in its surrounding neighborhood. But not all stars are the same as the Sun. Some are hotter and brighter, while many more are cooler and fainter. Inevitably we tend to see the bright ones rather than the faint ones, and while it is often said that the Sun is an average star, this does not mean to say that it is a typical star. As many as 90 percent of all stars are fainter than the Sun. Furthermore, fewer than half the points of light we see in the sky are single stars: about 55 percent are double or multiple stars, with two or more companions. So the Sun is anything but ordinary, though it is roughly midway between the brightest and faintest, and also between the most and the least massive.

Stars form from hydrogen gas clouds within the galaxy. These clouds condense under the pull of gravity and begin to heat up. Eventually the central temperature becomes so high that nuclear reactions start, and the star truly lights up. These reactions convert hydrogen into helium, releasing a little energy as they do so; eventually, heavier elements such as nitrogen and carbon are formed, and these also participate in the reactions. It is only within stars that temperatures are high enough to create all the heavier elements that we ourselves are composed of, as well as the rest of the physical world.

By human standards, stars live for a very long time, so the starry sky that we see today is essentially the same as that seen by the very earliest human eyes a few hundred thousand years ago. Stars take millions of years to evolve. A star's future is pretty well mapped out for it as soon as it forms, because its subsequent evolution depends largely on its initial mass. The most massive stars, which may be up to 100 times the mass of the Sun (though this figure is constantly debated), are also the brightest and the most spendthrift of their resources. They last a matter of only a few million years before swelling to become blue or red supergiant stars and then exploding as supernovae, leaving behind tiny neutron stars, pulsars or even black holes as remnants. These objects are not observable by amateurs, even though the catalogs of Go To telescopes may include them.

A star like the Sun, however, has a normal lifetime of about 10 billion years. Near the end of its lifetime it swells and its surface cools, and it becomes what is known as a red giant. It then blows off its outer layers until the core has a small diameter and a low light output, though a higher surface temperature – at which point it is a white dwarf.

As far as stars are concerned, the terms "giant" and "dwarf" usually refer solely to the stage in a star's life rather than to its actual mass or diameter. So even a star much more massive than the Sun may still be a dwarf star until it expands near the end of its life and becomes either a blue giant or a red giant. A star of three-quarters of the Sun's mass has maybe a tenth of its brightness and is known as a red dwarf. It will live twice as long as the Sun before becoming a red giant. (The dimmest red dwarfs have lives of several hundred billion years.)

The hotter the star, the bluer its color. This is the same color range as that of any glowing or incandescent object, such as the filament of an ordinary light bulb. A dim 40-watt bulb gives out a very much redder light than a 100-watt bulb. The reddest stars are comparable in color with a 100-watt light bulb, while sunlight is pure white, and stars hotter than the Sun appear distinctly bluish in color. As our eyes readily adapt to regard whatever is the ambient illumination as white, these colors are usually best seen by comparison. The best known is the difference between Betelgeuse, a red supergiant star, and Rigel, a blue supergiant star. The two make a striking contrast as they are both within the constellation of Orion. The contrasting colors of some double stars are obvious because they are close together, even though individually they might appear white.

Star types are described by a sequence of letters, often referred to in observing manuals. At one time the sequence was in alphabetical order, but now the reasons for classification have changed and the sequence runs O, B, A, F, G, K, M. The hottest, bluest and brightest stars are types O and B, while the A, F, and G stars are essentially white. Types K and M stars are red.

◀ *Betelgeuse (top left) and Rigel (lower right) in Orion are stars of strongly contrasting colors. A filter has been used to spread the star images and make the colors more obvious.*

Examples of stars of each type:

O Naos (Zeta Puppis), Zeta Ophiuchi
B Alnilam (Epsilon Orionis), Rigel (Beta Orionis), Spica (Alpha Virginis)
A Sirius (Alpha Canis Majoris), Vega (Alpha Lyrae)
F Canopus (Alpha Carinae), Altair (Alpha Aquilae)
G Sun, Alpha Centauri, Capella (Alpha Aurigae)
K Arcturus (Alpha Boötis), Pollux (Beta Geminorum)
M Betelgeuse (Alpha Orionis), Antares (Beta Scorpii)

Star details

Astronomers have painstakingly established distances, masses, true luminosities, compositions, and ages for large numbers of stars. Absolute certainties are rare in astrophysics, however. An accuracy of 10 or 20 percent is regarded as good, while the distances and other details of most stars are known to only about a factor of 2. The distances of stars are crucial to getting accurate information about them. These are usually measured by the parallax method – observing the slight shift in a star's position against the starry background as seen from opposite sides of the Earth's orbit. This is a reasonably accurate method for stars within a few hundred light years of Earth.

Double stars provide a vital source of information on star masses. Many stars are not single but instead are in pairs in orbit around each other, and the orbital period depends on the distance between the stars and their masses. If the stars are close enough that their distances from Earth can be measured by parallax, their masses can be estimated. In this way, astronomers can establish the masses of a range of stars of different types. If the distance of a star is known, its true brightness can be calculated from its apparent brightness in the sky.

It is then a matter of assuming that other stars of the same type are similar in their characteristics to those that are known, but there are many uncertainties. The distances to nearby galaxies can be estimated by looking for stars that look similar to known stars, and in this way the distance scale of the entire Universe is built up.

Observing stars

Although the word "astronomer" means "star namer," amateur astronomers probably spend less time observing stars than any other type of object. Individual stars in particular are usually mere stepping stones to other, more interesting objects. There is little difference in appearance between one star and another other than its brightness. In terms of visual spectacle an ordinary star does not have the same appeal as a planet, say. However, there are some things to be learned from studying even individual stars. They can be used to test a telescope or the quality of the seeing (see pages 20–21), for example, and the colors of stars tell you something about them.

To all intents and purposes stars are simply points of light. Though many of them are larger than the Sun, they appear as points of light because they are so far away. The distances to stars are usually given not in kilometers but light years (see page 13). Consider the Sun, our nearest star. Its light, traveling at 300,000 km/s, takes just 8 minutes to reach the Earth from its distance of 149 million kilometers. If we were to travel away from the Sun at the speed of light, within a month it would appear so small that even a 250 mm telescope would no longer show it as a disk. To reach the nearest star, however, we would have to travel for over four *years* at the speed of light. This star, Proxima Centauri, is 4.28 light years away.

▲ *The Hubble Space Telescope has taken one of the few photographs of the actual disk of a star – in this case the red supergiant Betelgeuse, which has a disk less than 0.1 arc seconds across.*

So if the Sun were as distant as even the nearest star it would be far too tiny for an amateur telescope to see as anything other than a point of light. The disks that you see on photographs of stars are caused by the spreading of light on the image-sensing medium used to take the picture, and have nothing to do with the actual diameters of the stars. Not even the Hubble Space Telescope has the power to show the nearest star as a disk. In fact, it can reveal the disks of only one or two of the very largest nearby stars, and these are rare exceptions.

View a star through a telescope and it does not actually appear point-like. There are two main reasons for this – the resolving power of the telescope (see page 21) and the seeing. A small telescope, which has limited resolving power, will show a star as a tiny disk, known as the *Airy Disk*. The size of this disk is set by the wave properties of light. In the case of a 50 mm telescope, the Airy Disk is easily visible with a magnification of 100. The image of an object that is not a point of light can be thought of as being composed of large numbers of overlapping Airy Disks, one from each point on the object. The larger the telescope the smaller the Airy Disk, so the image from a larger telescope is effectively composed of finer dots. Unlike an image in a book, however, all the dots merge into each other rather than being seen separately.

The quality of the atmospheric seeing also affects the size of a star image. Unless the atmosphere is totally still, which is never the case, the effect of seeing is to enlarge the star image. This seeing disk, as it is called, is constantly on the move, and

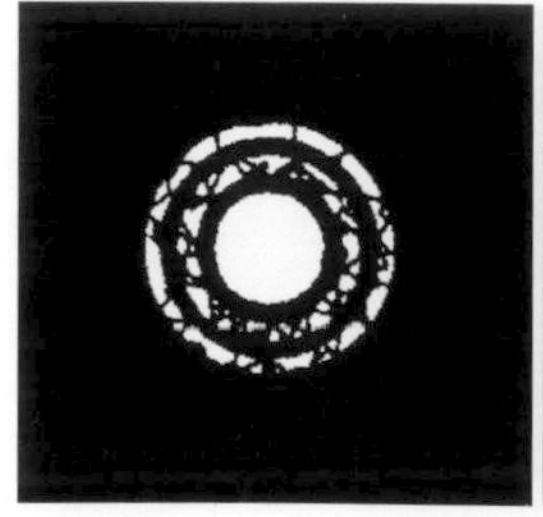

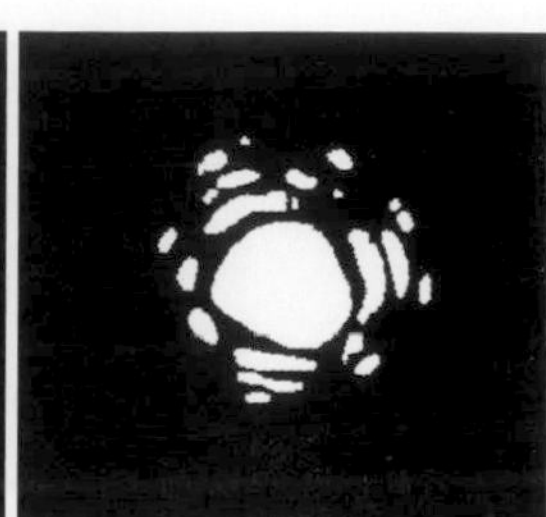
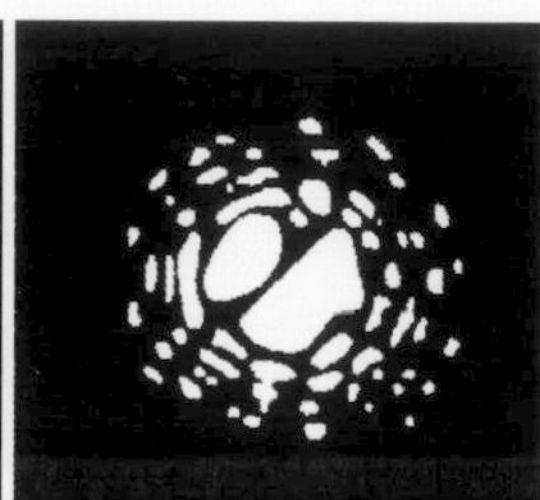
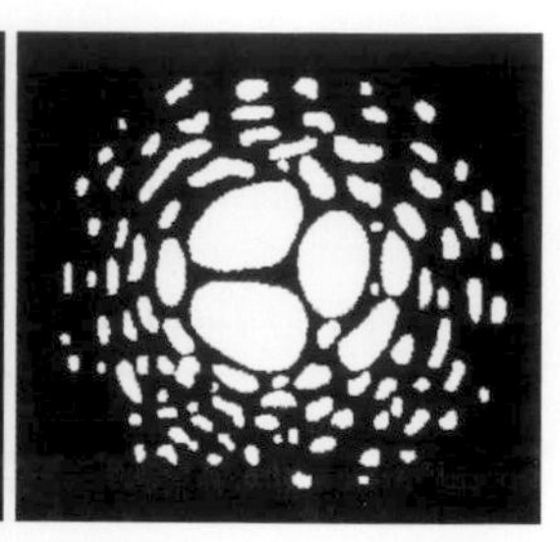

▲ *Increasingly poor atmospheric seeing degrades a star image from a perfect Airy Disk (left) to a larger and less well-defined blur. The rings seen around the Airy Disk are very subtle and are known as diffraction rings.*

under extreme circumstances it can look like a sparkling writhing amoeba, with projections appearing and disappearing as you watch. With a telescope larger than about 150 mm aperture, the Airy Disk is often smaller than the smearing effect of the seeing disk and is not easily visible. Typically, fairly good seeing produces disks about 1 arc second across. Two or 3 arc seconds is not unusual, while a quarter of an arc second is particularly good. Visual observers learn to make the most of the odd moment of better seeing.

The quality of the telescope can also affect the size and shape of a star image's disk. Triangular or other noncircular images are caused by poorly manufactured optics, or optics that are pinched in their mounting cells, causing them to be distorted. Sources of local heat, such as uneven cooling of the telescope components or even someone standing right below the line of sight, can produce distorted, moving images.

So observing a star image can tell you a great deal about the telescope and the atmospheric seeing conditions, though not very much about the star itself. The main details that you can find out about a star are its brightness and whether it has a companion star (in which case it is known as a double or binary star). Its color may reveal its star type; this is particularly the case with the brighter stars because their colors are easier to see.

Because stars have point-like images, it is possible to beat light pollution to a certain extent. Magnify a point of light and it remains a point of light of the same brightness, but the background will become darker with increased magnification because the light is spread over a larger area. So you can improve the contrast between a star and the background by using a higher power. In practice, the seeing can enlarge the star image as well, but the technique is useful when observing any stars or star clusters.

Double stars

Most double stars lose out in the popularity stakes to nebulae, clusters, and galaxies. Through the telescope you just see two stars side by side, often with one much fainter than the other so that it is hard to spot, and you might be forgiven for feeling, after a while, that if you have seen one double star you have seen the lot. But there are exceptions – some doubles are famous for their contrasting colors, while others have several components, which makes them a more interesting sight. Tracking down double stars is a good occupation for a night when the seeing is good but the transparency (see page 21) is not perfect.

True double stars consist of a pair of stars that are physically close to each other – usually within a distance comparable with those of the planets round the Sun. They mutually orbit each other with periods that range from days up to thousands of years. Those stars that orbit within days are so close that they cannot be seen separately, and can only be observed using special techniques. There are also some stars that appear double, but where the two stars just happen to be in the same line of sight. These are known as *optical doubles*.

Although some amateurs make detailed measurements of double stars with a view to improving our scientific knowledge, most are content simply to view them from time to time. Observing a double star that is close to the resolving power of your telescope is a good way to confirm its quality, and it also tests your own observing ability at high magnification.

The important features of a double star to consider are the brightnesses of the two components and their separation. The easiest doubles are those where the two stars are of more

► *The star Izar, Epsilon Boötis, is a double with a separation of 2.8 arc seconds. The main star is of type K and has a magnitude of 2.4. The A-type secondary star is at magnitude 4.9. Izar is a favorite among double-star observers because of its attractive contrasting colors.*

or less equal brightness and have a large separation compared with the resolving power of your telescope. As the magnitude difference between the two increases, or the separation decreases, the doubles become more tricky. A difference of three or four magnitudes is enough to make a star with well-separated components hard to see except at high magnification. Separations are usually measured in arc seconds. Another measurement that is usually given is the *position angle*. This refers to the angle in the sky of the line from the brighter to the fainter star, with north being 0° and east being 90°. The scale therefore goes counterclockwise. The brighter star is usually referred to as A, the secondary B, and subsequent components of a multiple star C and so on.

Variable stars

Some stars are variable in brightness, for one reason or another. In some cases – known as *eclipsing binaries* – the orbital plane of a double star happens to be edge-on to the Earth, so we see mutual eclipses of the stars. Algol in Perseus is the best-known example of this type. More often, the variations result from a single star pulsating in brightness because it has reached an unstable stage toward the end of its life. Some of these variable stars are known as *Cepheids*, after the first star of this type to be identified, Delta Cephei. These stars vary in brightness over a matter of days or even weeks, with a characteristic slow fall and more rapid rise. They have the important property that the longer the period of variation, the brighter the star. So if a star is seen to vary in this way, its true brightness is known

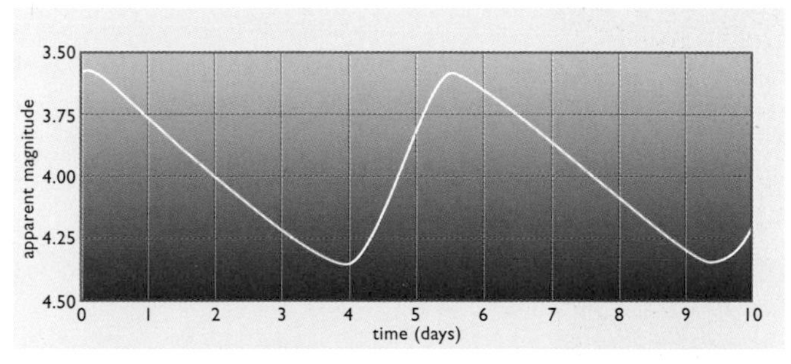

▲ *The light variations of Delta Cephei, which pulsates regularly with a period of 5 days 9 hours, are easily followed with the naked eye. A comparison chart is given on page 88.*

once its period has been measured, and hence its distance can be found from its apparent brightness.

There are many other types of variable star, some of which vary regularly, while others are unpredictable. The variations in the irregular types are often caused by pulsations in an old star, but others, known as *eruptive variables*, are subject to flickers or flares. Many eruptive variables are young stars, but flares are also a feature of some red dwarf stars. Another cause of variability is the rotation of a star with numerous spots or brighter regions.

Cataclysmic changes generally occur when double stars are so close together that material falls from one on to the surface of the other, causing an explosive outburst, just as when fat drops on to a barbecue. The most dramatic and best known of these stars are *novae*. A nova is a star that appears apparently out of nowhere, and in rare cases it may for a few weeks be one of the brightest stars in the sky. In fact a nova is a faint star whose close companion star has started to evolve into a red giant, resulting in a tremendous flare-up. Most novae are fairly faint, but occasionally one reaches naked-eye visibility.

Making variable-star estimates

The eye is quite good at deciding whether one star is brighter than another, and by how much. So it is possible to make reasonably accurate estimates of the brightness of a variable star by comparing it with nearby reference stars of known brightness. Ideally there would be a good range of stars of similar brightness within the same field of view as the variable star, making it easy to decide which one it is equal to. In practice, however, the comparison stars are usually more widely spaced in brightness, and may not be in the same field of view. Typically you have to estimate where in the brightness gap between the two comparisons the variable lies – say two-fifths or three-quarters of the way between them. With experience, observers can recognize a brightness difference of a fraction of a magnitude, and can estimate the brightness of a star that is brighter or fainter than the comparisons.

The comparison stars used when making estimates of variables must be chosen with care, avoiding stars with strong colors or that may themselves be variable, and ensuring that the stars have accurately determined magnitudes. So it is best to make estimates using comparison charts published by organizations such as the American Association of Variable Star Observers or the British Astronomical Association's Variable Star Section. Mostly these charts are for stars that require a telescope or at least binoculars, but there are a few that can be estimated using the naked eye alone.

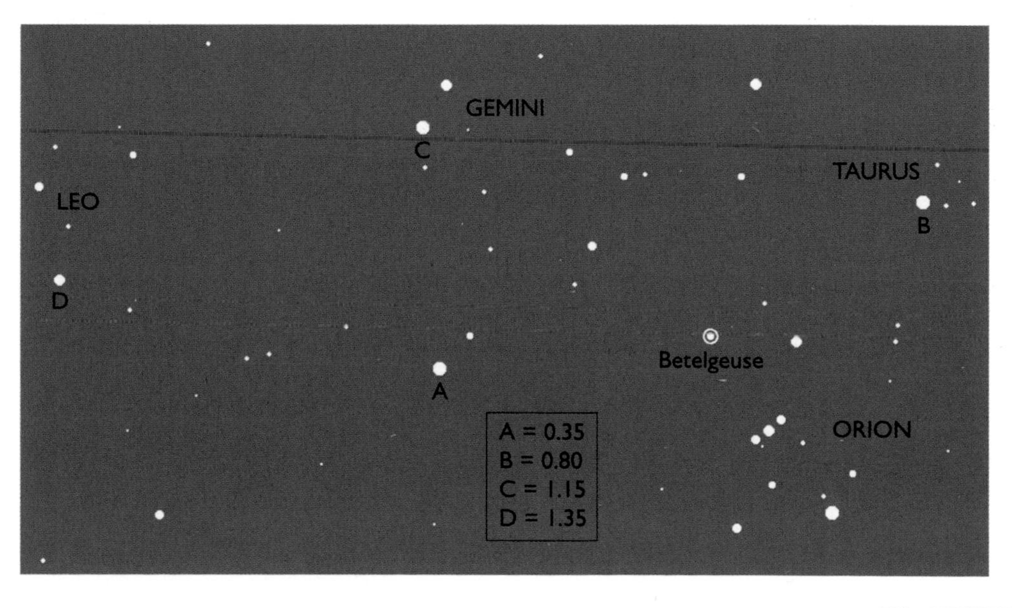

► *A comparison chart for estimating the magnitude of the irregular variable star Betelgeuse in Orion. Decide whether the star is brighter or fainter than the comparison stars shown, and by how much. Then use the comparison star magnitudes to estimate the magnitude of Betelgeuse.*

Star clusters

Stars are often born together and in some cases they may remain in a group for a considerable period of time. Star clusters are useful to professional astronomers because they contain stars that share a common origin and are clearly all at roughly the same distance. To amateur astronomers they are appealing because of their beauty. A field of view full of stars is a sight to behold, and the occasional presence of a strongly colored star or two adds to the effect. Many clusters are easy to spot even from light-polluted skies which blot out most diffuse nebulae and galaxies.

There are two general types of cluster – the loose or *open cluster*, and the *globular cluster*. Open clusters range from a few scattered stars to a group of several hundreds all within a fairly small region of sky. All the stars are kept in a compact group by their mutual gravitational forces, though over a period of tens of millions of years they may slowly spread apart and eventually separate. They are found within the spiral arms of our Galaxy, so they are mostly seen along the line of the Milky Way, and may contain quite young stars.

Globular clusters, however, consist of thousands or even hundreds of thousands of stars in a generally spherical body. Their distribution is also quite different – they surround the nucleus of the galaxy but at a considerable distance from it. Most of the stars are old, and there is no trace of gas or dust within the cluster. Although many globular clusters happen to lie in the same line of sight as parts of the Milky Way, there are plenty outside the general plane of the Milky Way. Because they surround the nucleus, they are a particular feature of the evening skies of the middle of the year, when the center of the Galaxy is prominent. By contrast, there are virtually no globulars in December skies.

▼ The Pleiades cluster, M45, is the brightest and best-known star cluster, and is visible to the naked eye. It can be seen in all conditions, and binoculars show it particularly well.

Some nearby open clusters are easily visible with the naked eye, the Pleiades and the Coma Berenices star cluster being good examples. Binoculars improve their appearance, but many telescopes have too small a field of view to do them justice. Binoculars bring many other clusters within range, such as M35 in Gemini or M25 in Sagittarius. Though these may be visible with the naked eye under good conditions, binoculars reveal individual stars and low powers on telescopes reveal a glittering array of stars. Yet other clusters such as the Wild Duck Cluster and the Jewel Box or Kappa Crucis Cluster are too small to be resolved into stars with binoculars, so telescopes with medium powers are really needed to show the cluster well.

With the naked eye or small instruments, many star clusters are indistinguishable from the other deep sky objects, such as nebulae and galaxies, because the stars in them are too faint to be seen individually. Globular clusters in particular can also look like distant comets. In the 18th century, a French astronomer named Charles Messier listed many such objects that he encountered while searching for comets, so that he could eliminate them quickly from his searches. His listing remains in use, and many of the brighter objects are known by their Messier numbers, either as, for example, Messier 35 or just plain M35.

The brighter globular clusters are visible with binoculars (and in some cases to the naked eye) as hazy stars, but are best seen with telescopes. The brighter ones, such as Omega Centauri or M22, are resolved into stars with low powers and are a spectacular sight through a medium-sized telescope with a power of about 50 or 75. The fainter globulars require a similar or greater magnification to reveal them properly, and may be hard to pick up at all in light-polluted skies with small telescopes.

Photographs usually make the cores of globular clusters appear brilliant, with all the stars combining together. This is not the case through a telescope, however, and you can always pick out individual stars even at the center of the brightest globular clusters, though there may be a background glow of stars that are too faint to see. This is because photographs enlarge star images, whereas your eye sees them as points of light of increasing brightness. Few photographs do justice to the appearance of a great globular cluster in a good sky.

◀ A drawing of globular cluster M3 in Canes Venatici made using a 150 mm reflector with a magnification of 150.

▲ *The globular cluster M92, in the constellation of Hercules, is about 28,000 light years away. The cluster is about 100 light years across and contains several hundred thousand stars.*

Nebulae

The word *nebula* is Latin for "cloud." The ones referred to are not those that ruin our view of the sky, but those that appear as misty patches when seen through a telescope. The plural is *nebulae*, pronounced "neb-you-lee."

There are many small cloudy patches in the sky, and at one time all were referred to as nebulae. By the mid-20th century, however, their true physical nature had become clear and now those that are actually galaxies are referred to as such. True nebulae are gas clouds of one type or another, though at the telescope it is hard to tell one type from another just by looking.

There are two basic types of nebula, one occurring at the beginning of a star's life, and the other at the end. The former are known as diffuse nebulae, and they are clouds of gas and dust – the raw material from which stars form. They can be either bright (*H II regions* or *reflection nebulae*) or *dark nebulae*. The latter are *planetary nebulae* or *supernova remnants*, and they are shells of gas thrown off at the end of a star's life. Some are small and still surround the dying star, while others may cover a wide area of sky and may be the remains of an exploded star.

Gas and dust nebulae

Our Galaxy contains vast reserves of material from which stars can form. This is mostly hydrogen gas, but there are many other constituents mixed in, notably molecules and dust. Such clouds are mostly opaque, and they are responsible for huge dark nebulae such as the Coalsack near the Southern Cross, and the Cygnus Rift in the Milky Way from Cygnus to Sagittarius. Dark nebulae are usually only visible in silhouette against brighter objects, such as star clouds or bright nebulae, and few of them are visible with small telescopes.

A star near a cloud of gas or dust may provide some illumination. The main way in which this occurs is a form of fluorescence – the gas emits light of its own when it receives a blast of ultraviolet light from a nearby hot star, often one that has recently formed from the gas cloud itself. Hydrogen glows pink, but in some cases oxygen and nitrogen may glow green. Nebulae of

► *The region around the Cone Nebula in the constellation of Monoceros illustrates all three types of gaseous nebula. The pink nebulae are H II regions, while the dark intrusion known as the Cone, at left, is a dark nebula. The lack of stars at the left-hand edge indicates that the dark nebula extends throughout this area and is hiding more distant stars. Within the cluster NGC 2264 at right are dust clouds shining as blue reflection nebulae; they are illuminated by nearby stars.*

this sort are called H II regions – pronounced H-two – from the nature of the hydrogen molecule of which they mostly consist.

The Galaxy is a dirty place, as a result of soot produced by stars. This may sound bizarre, given that stars and indeed the Sun do not burn conventionally, but instead shine as a result of nuclear reactions deep within them. These reactions, though, consume hydrogen, the simplest element, and in the process create more complex elements, such as carbon and nitrogen. During a star's lifetime, and particularly when it reaches the end of its life, it throws off shells of these materials, which form sheets of dust. The carbon grains gain a coating of water ice – which is, of course, nothing more than a combination of hydrogen and oxygen. So ice and soot are major constituents of the material between the stars.

Starlight shining on this material illuminates it, and just as the light scattered by woodsmoke appears blue, so does the illuminated dust in space. These nebulae are less common than H II regions, and are known as reflection nebulae.

Planetary nebulae

A star like the Sun shines steadily for billions of years but eventually its reserves of hydrogen can no longer sustain it, and it begins to shed material into space. Gas expands away from the star, creating a huge shell surrounding the star that in some cases is visible through a small telescope. These shells often appear as tiny disks about the same size and brightness as one of the distant planets, Uranus or Neptune. For this reason they are referred to as planetary nebulae.

These objects are much more compact than H II or reflection nebulae, and sometimes show a distinct blue-green color when seen through a telescope. Photographs taken with the Hubble Space Telescope, for example, show beautiful colored rings or tubes of gas. Usually the naked eye is not sensitive enough to see color in nebulae. However, though most planetary nebulae are fairly faint they are usually quite compact, and their surface brightness is quite high. When looking for color, it is the surface brightness rather than the overall brightness that counts, so in a medium-sized telescope, say 150 to 200 mm, on a good night, the color can be easily visible.

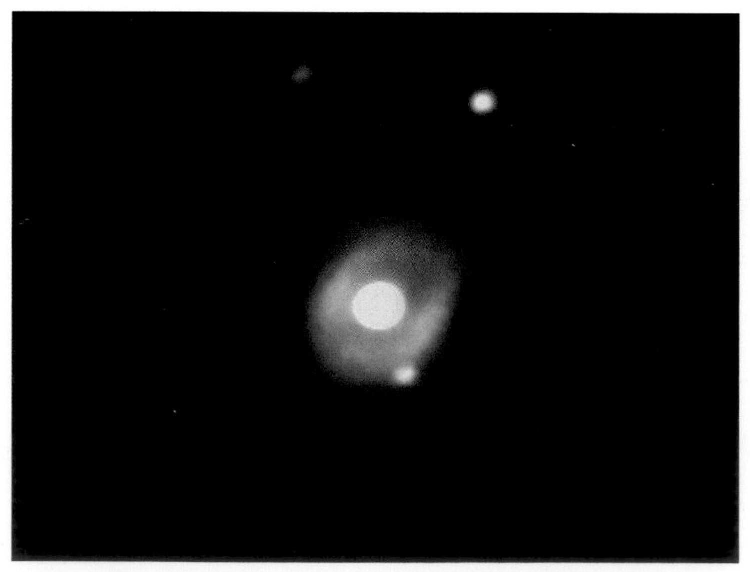

▲ *One of the brightest planetary nebulae, NGC 3132 or the Eight-Burst Nebula. It lies in the southern constellation of Vela.*

Supernova remnants

Most stars die a fairly slow death, puffing off shells of gas and finally dwindling away as a dim star known as a white dwarf. The most massive stars, however, die spectacularly in a vast explosion which can briefly be as bright as all the other stars in the Galaxy put together. This event is known as a supernova, and it results in very large expanding shells of gas which collide with the material between the stars, causing it to glow pink or blue. These are supernova remnants, and a few are visible or can be photographed using cameras mounted on driven telescopes.

The light from nebulae

Many nebulae are quite faint, but fortunately there is one major difference between their light and that of other celestial objects, such as stars. To see the difference, astronomers use the technique of *spectroscopy*. This involves splitting the light from the object into its component colors, using either a prism or a diffraction grating (a flat surface with very fine grooves). We are all familiar with CDs or reflective paper that is embossed with fine pits or lines; they split light in exactly the same way.

White light from the Sun, stars, and glowing light sources such as light bulbs consists of a rainbow of all colors from deep violet (short wavelengths of light) to deep red (long wavelengths). This is known as *continuous emission*. A reddish star has more red than blue light, and a bluish star vice versa, but they both contain all the colors of the rainbow.

A nebula, however, usually emits only very specific colors of light, and this is known as *line emission*. Hydrogen, for example, emits a few individual colors or wavelengths of light. The brightest is deep red, with the next brightest being green and the next being blue. These colors combine to make a hydrogen nebula appear pink. Our eyes are not very sensitive to red light, so visual observers tend to see only the green and blue colors, but film in particular is more sensitive to the red, which is why it is possible to photograph some nebulae that are virtually undetectable to the eye. This is just about the only remaining advantage of using film for astrophotography.

Planetary nebulae and supernova remnants also have line emission, though often of different colors. Only reflection nebulae have continuous emission.

The importance of all this is that when we are trying to observe nebulae we are usually looking through a certain amount of light pollution. While some sources of light pollution consist of continuous emission, many have either line emission or a strong color bias. This makes it possible to manufacture filters that cut out some of the light pollution (known as light-pollution rejection or LPR filters), while allowing through the general bands of color that are emitted by nebulae. These are broadband filters, but you can also get narrowband filters, which allow through only the light from the nebula. These are known by the type of gas that emits those colors. An O III filter, for example, allows through only the light from oxygen while blocking everything else.

Observing diffuse nebulae

Some H II regions are bright enough to be visible even in poor skies. The very brightest, such as the Orion Nebula and the Eta Carinae Nebula, are visible to the naked eye in a fairly dark

► *The Veil Nebula in Cygnus is a good example of a supernova remnant. Its circular shell is over 100 light years in diameter. It has been expanding for more than 20,000 years from a central star which can now no longer be identified.*

▼ *A drawing of part of the Veil Nebula in Cygnus, made using a 220 mm telescope and an O III nebula filter.*

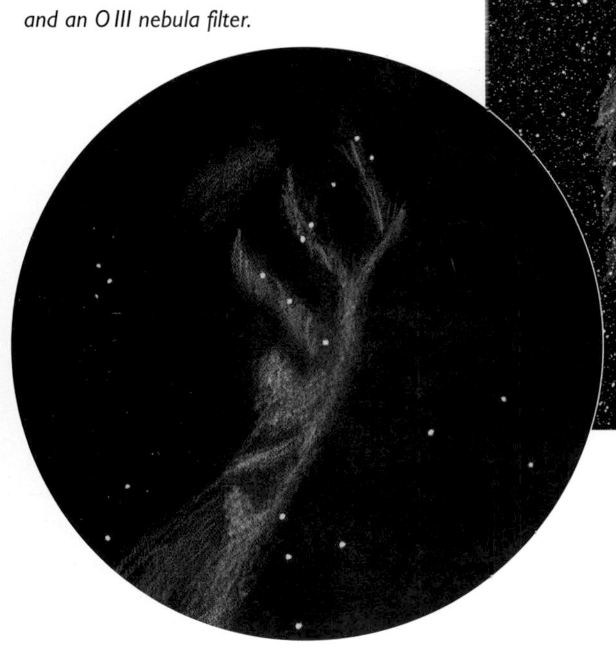

sky and with binoculars even in a light-polluted sky. For the best views, however, the darker the conditions, the better.

There is a wide range of sizes and brightnesses of nebulae, so different approaches are needed depending on the object. While a bright and large nebula such as the Lagoon Nebula, M8, is easily visible in binoculars or a small telescope, other large nebulae, such as the North America Nebula or the Rosette Nebula, are large and faint. Though they may require eagle eyes and perfect skies to be seen visually, they can be photographed quite easily with time exposures using ordinary cameras mounted piggyback on driven telescopes. Film and CCDs are more sensitive to the red light of hydrogen than is the eye, so they can pick out large nebulae from the light pollution, even from suburban skies. Most digital cameras, however, include an IR filter which reduces their red sensitivity.

The smaller and fainter nebulae require a telescope, usually at low power, and clear skies. This is an occasion where a specialized filter is worth trying on the eyepiece. Simple color filters are not generally of much use for nebulae, as their effect is too general – they cut down the light of the nebula as much as they do the sky background. But LPR filters or specialized narrowband filters such as an O III filter can help, though because they absorb light they are unsuitable for use with small telescopes.

Don't expect too much from LPR filters – they cannot magically transform a city sky into a country sky. But they can make gaseous nebulae more visible, even in what would normally be regarded as very dark skies. Rarely do you find a site where there is no background light at all. Narrowband filters will not cope with really bad light pollution, but they can have a magical effect where the object in question is on the borderline of visibility.

Galaxies

There are literally millions of galaxies out there to be observed, and a surprising number are visible with quite small telescopes under good conditions. It has to be said that the majority of galaxies are not spectacular, but they do have a great appeal. These are vast star systems like our own Milky Way Galaxy, though incredibly remote. Those that you can see with small telescopes are among the nearest, yet you can gaze at objects so distant that their light left them while dinosaurs were still walking the Earth. Each one undoubtedly has its own great civilizations, and alien eyes are looking out from them across the Universe to our own Galaxy, which is as remote to them as they are to us. Even in the fantasy world of *Star Trek*, ordinary travel between galaxies is out of the question.

Distances to galaxies are measured in millions of light years. Even the nearest large galaxy, the Andromeda Galaxy, is about 2.5 million light years away. Some of the brighter ones that you can easily see with binoculars, such as M81, are around 12 million light years away, while the galaxies of the Virgo Cluster are about 50 million light years distant, and are at the limit of binoculars and the smallest telescopes. With larger apertures you can see numerous galaxies that are 200 million or more light years from us. Such distances are not well-determined though, so it is hard to put a figure on the most distant galaxy you can see with any particular instrument.

Although all galaxies are great collections of stars, they fall into various categories as a result of their appearance. The main types visible with amateur telescopes are *elliptical* and *spiral*, but there are also *irregular* and *lenticular* galaxies. The most common throughout the Universe are irregulars, but they tend to be smaller and not so easily visible with amateur telescopes. Exceptions are the two Magellanic Clouds visible in the southern hemisphere, which are only visible because they are close, being companions to our own Galaxy.

Elliptical galaxies include the largest of all galaxies, though they come in all sizes. They are, as the name implies, elliptical

▲ *The brightest elliptical galaxy, M32, is a companion to the Andromeda Galaxy, M31. Though only a dwarf elliptical galaxy, its closeness makes it is easy to see.*

in shape. They generally have a very smooth gradation of brightness from the center to the edge with no internal structure visible in amateur telescopes, except sometimes an almost starlike nucleus. Some appear virtually spherical, while others are elongated. It is difficult to tell the true shape of an elliptical galaxy just by looking at it. If it appears in the telescope as a sphere, it could be a genuinely spherical galaxy, or it could be a flattened sphere (like a tangerine) seen end-on, or it could even have three axes of different lengths, such as a flattened American football or rugby ball, seen at such an angle that it appears spherical. Other elliptical galaxies, however, do appear elongated.

Elliptical galaxies contain virtually no gas or dust, and they appear yellowish or even reddish in photographs because many of their stars are old rather than young.

Spiral galaxies are undoubtedly the most popular among observers, as they have the classic and beautiful spiral arms. The nucleus of a spiral galaxy is virtually indistinguishable from an elliptical galaxy and has the same yellowish color. However, there is a flattened disk through its center which displays some sort of spiral structure. The disk contains gas and dust, and has bright young blue stars as well as pink H II regions when photographed in detail. Some galaxies have a nucleus that is bar-shaped rather than spherical, in which case they are known as *barred spiral galaxies*. Lenticular galaxies are elliptical, but they have a disk that appears to have no obvious spiral structure.

Although photographs of spiral galaxies may appear to show individual stars, we only ever see the general mass of stars rather than individual ones. Occasionally there may be a massive cluster of stars that looks like a single star, but in fact even the Hubble Space Telescope can only just distinguish individual stars in most galaxies. The only exception is when a star in a galaxy explodes in a supernova explosion, in which case it may briefly appear as a star within or near to the galaxy. Some amateur astronomers spend many hours searching for these events, usually using CCD cameras on automated telescopes, as supernovae are of great interest to professional astronomers. Bear in mind that most of the stars seen in the field of view of a galaxy are much closer ordinary stars within our own Galaxy.

Observing galaxies

Apart from the Andromeda Galaxy and the two Magellanic Clouds, which are visible with the naked eye, most galaxies are rather small and faint. The brighter ones are visible with binoculars, but most require a telescope and clear, dark conditions. As with all deep sky objects, the larger the aperture and the darker your skies, the better the view you will get. But under the right conditions even a 110 mm telescope will reveal hundreds of galaxies.

Being quite small, most galaxies take some finding unless you have a Go To telescope. It helps if there are some bright stars fairly near to aid star-hopping, but in the case of the Virgo Cluster, which contains a large number of galaxies, there are few stars in the area. The galaxies might appear in the telescope, but it is hard to tell which you are looking at. The trick is to choose an easily repeatable route into the cluster, as described on page 115, and work from there.

Once you have found a galaxy, notice its shape and any stars nearby. Spiral galaxies are often not obvious even though their photographs show spectacular arms. With a small telescope, frequently all you can see is the nucleus of the spiral, which looks like an elliptical galaxy. Only in good conditions and with practice can you see the spiral arms, much fainter and well outside the nucleus. The Andromeda Galaxy, M31, is a case in point. It is visible with the naked eye by averted vision in

► *An edge-on spiral galaxy, M74, in Pisces.*

◀ *M109 in Ursa Major is a good example of a barred spiral galaxy.*

average skies, and with binoculars even from city centers. But what you usually see is the nucleus only. In dark skies the galaxy extends way beyond this, and structure becomes visible, notably the sharp edge to the nearer spiral arm, which is caused by a line of dust clouds. Some galaxies are spindle-shaped, and edge-on spirals in particular can be almost needle-like in their appearance.

Look for subtle variations in the brightness of the galaxy, and give your eye time to get used to the field of view. Although in theory the eye cannot build up light over time like a CCD, some observers claim that they can see more the longer they keep looking at an object.

Narrowband filters suitable for observing nebulae will not be of any use with galaxies, as the light from a galaxy is essentially white light rather than line emission from a nebula. Light-pollution filters, however, can be of value, though the effectiveness of a particular brand of filter will depend on your local conditions and the particular sources of light pollution. City centers have a predominance of white light, with essentially the same color distribution as the galaxy, so using a filter will simply cut down the light from both sources. But in suburban areas with low-pressure sodium (orange) or mercury (blue-white) lighting, an LPR filter could help.

The magnitude of a galaxy quoted in catalogs is not a very good guide to its visibility, as this value gives the total light output from a considerable area. The appearance is quite different from that of a star of the same brightness, whose light is concentrated into a point. Also, face-on spirals are more difficult to observe than edge-on spirals. This is because the light from the latter is contained within a narrow needle shape rather than spread over a larger area, as in the case of a face-on spiral.

Observing deep sky objects

Observing deep sky objects, whether clusters, nebulae or galaxies, requires quite different techniques from those used for objects within the Solar System. Deep sky objects are often quite faint and harder to find. Light pollution is frequently a major problem. While urban observers may struggle to see a comparatively large and bright object such as the Crab Nebula, those in dark areas may face similar problems trying to find much fainter objects. Just finding the object is a challenge in itself. If you can't see it, is the object too faint for your telescope, or is it too small or too large for the magnification you are using, or is it hidden in the light pollution – or are you looking in the wrong place to start with? Owners of the smaller Go To telescopes need not feel smug, because although they may have been pointed in the right direction, all the other questions still apply.

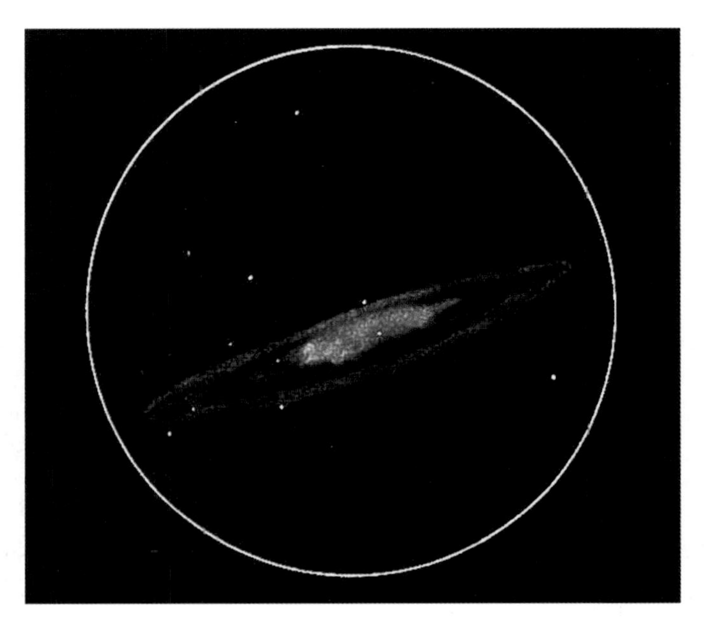

▲ *A drawing of the spiral galaxy NGC 253 in Sculptor, made using a 500 mm Dobsonian reflector.*

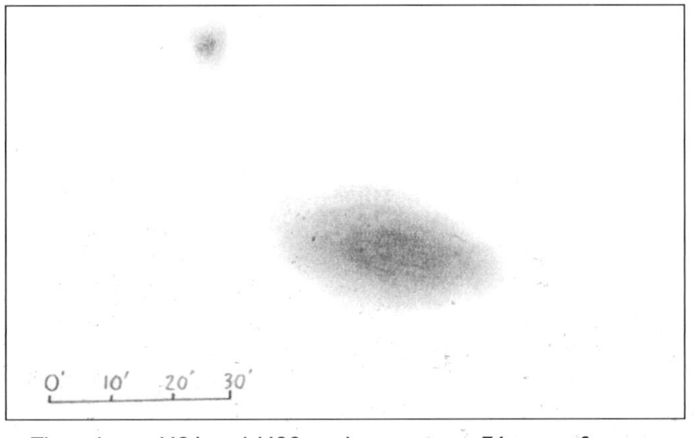

▲ *The galaxies M31 and M32 as drawn using a 76 mm refractor. The larger object is M31. Compare its extent with the photograph on page 82, in which M32 is the small circular fuzzy object below the nucleus of M31. The scale is in arc minutes.*

Observing tips

All deep sky objects by their nature are best viewed in a very dark sky with a complete absence of extraneous light and with a telescope in perfect condition. These are criteria that rarely fully apply, but you can do your best to make the most of what you have.

Make sure that your telescope is up to the job. Clean optics make a lot of difference, but it is better to put up with a little dust than to risk scratching a delicate surface. Keep covers on optics until they are needed and try to avoid dewing up. If dew starts to form on a lens or corrector plate, cover it straight away or blow warm air on it gently to dry it off. A portable hairdryer is a useful accessory if you have a 12-volt power supply handy. Dust falling on dewed-up optics is likely to stick.

Discussion of the cleaning of optics is beyond the scope of this Atlas; suffice to say that it should only be carried out with care and when strictly necessary. Eyepieces as well as objectives and mirrors should be kept dry and dust-free.

Any dust on the optics will reduce the contrast between the object you are looking for and the sky background. Other causes of low contrast are poor baffling and reflections within the telescope tube. These are not easily altered with commercial instruments, but you can at least make sure that a refractor or SCT has a good dew cap, which should be matt black on the inside. In the case of a reflector, any worn black paint or shiny parts at the top of the tube should be attended to. But most importantly, try not to observe with extraneous light shining on the telescope or into your eyes. If possible move the telescope to a darker spot or set up a screen to block out the light – even a blanket on a clothesline or draped over a ladder will help.

Choose the correct eyepiece for the object you are trying to observe. Our observing notes give suggestions for the minimum power that gives a good view, but in general for diffuse nebulae you will need a low power, for galaxies a medium power, and for planetary nebulae a high power. Star clusters can be either low- or high-power objects depending on their size. The smaller star clusters and globular clusters generally require a fairly high power though they can usually be found with a lower power. Make sure you are comfortable when observing and that the eyepiece is at a convenient position.

Wait until the objects are as high in the sky as possible, and until conditions are as clear and dark as possible. Cold air is usually clearer than moist, warm air. Observing after midnight may help as some lighting does tend to be switched off then. Wait until the sky is properly dark – that is, the Sun is at least 12 degrees or, better still, 18 degrees below the horizon. In summer at high latitudes, such as in northern Europe, it never gets fully dark. Wait until the Moon is out of the way.

Avoid looking at any bright lights or TV or computer screens for up to an hour before observing, and use only a dim red light for looking at your star map. The maps in this Atlas are designed to be used with such a light. Practise using averted vision (see page 14). Tapping the telescope can help to reveal a tiny faint object – the eye is good at detecting movement.

▼ An amateur photograph of the Andromeda Galaxy, M31, from Wales, UK. It was taken with a CCD camera through a 200 mm reflecting telescope with a CCD autoguider attached to a separate telescope. Exposures were made through red, green, and blue filters with a total exposure time of 2 hours 30 minutes.

Finding deep sky objects

If you have an ordinary telescope with a finder, get used to the inverted image given by the finder and its field of view. The 5 × 24 finders on cheap telescopes are virtually useless for finding deep sky objects, and even the better quality 6 × 30 finders are of limited value where there is light pollution. Non-magnifying finders or sighting devices (see page 18) can be useful when you know roughly where the object you are searching for is to be found.

It is often easy to locate an object in binoculars yet hard to find the same object in the finder. This is because the finder generally inverts the view and has worse optical performance. The key to finding deep sky objects is either to have a good finder, or to know the performance of the finder you have so that you can allow for its deficiencies. Some observers draw a circle on a clear plastic overlay for their star map that shows the field of view of their finder, which can help when trying to recognize star patterns.

The most popular deep sky objects are usually those that are easy to find, either because they are bright or because they are conveniently located near to bright stars. Some comparatively bright objects can be overlooked simply because they are more tricky to locate. So the key to finding deep sky objects is to refine your finding skills. Go To telescopes can help greatly, but only if they are accurately set up in the first place. But without Go To, you need to develop your star-hopping skills.

Star-hopping involves starting from a star that you can find easily and jumping from there, one field of view at a time, to the object you want to find. If your finder shows few stars, you may have to use your lowest power and move the telescope one field of view at a time from a known star to the chosen object. An alternative method, if you have an equatorial mount reasonably accurately aligned, is to locate a star with the same or similar declination as the object you want, then sweep the telescope in right ascension only to find the object. If there is a star with the same right ascension as your chosen object, you can do the same but this time sweeping only in declination. Even if you have an altazimuth mount, this procedure will work when the object is close to the meridian.

All these methods require persistence and practice, but eventually you will find that you can pick out many deep sky objects quickly while owners of Go To telescopes are still fumbling with their key pads or changing their batteries. Nevertheless, there are times when all Dobsonian owners wish that they could have a bit of technical help in finding a particularly elusive object, so there is something to be said for both systems. Users of Dobsonians and other telescopes on altazimuth mounts must take into account that the axes of their telescopes are not usually aligned with the RA and Dec grid lines, so the orientation of the sky probably differs somewhat from the star map.

Drawing deep sky objects

Many observers like to make a drawing of what they see at the eyepiece as a simple and quick record of the object. A clipboard with a red light attached is a useful accessory, and most observers favor a selection of pencils of different densities. Sketches are invariably made in negative form, with dark shading and black stars, for simplicity, though it is easy enough these days to scan the drawing into a computer and invert the tones so as to get a more realistic result. It is usually necessary to emphasize the contrast of the view, because many objects are barely visible against the background. Making a drawing teaches you to look carefully at the object, and many people return to drawings they made with previous instruments, with younger eyes or in darker skies, to see how their perceptions have changed.

Star clusters are tricky to reproduce accurately, and many people settle for putting in the main stars and then stippling in the remainder, taking care to preserve the general appearance and shape of the object.

As with all astronomical observations, it is essential to make a note of the date, time, instrument, magnification, and seeing conditions.

Photographing deep sky objects

Although simple methods, using compact digital cameras and webcams, can give very good photographs of objects in the Solar System, most deep sky objects require a different level of complexity. Time exposures, often minutes or even hours in total duration, are the norm. Conventional webcams can easily pick up the brighter stars, and can give very good images of double stars, but that is about their limit without modification.

Successful deep sky photography requires a sensitive system and long exposures, which means using a well-aligned equatorial mount with a precise guiding system. Digital cameras, particularly digital SLR cameras which allow you to use long telephoto lenses, can be mounted piggyback on a driven telescope to obtain nice images of the larger and brighter deep sky objects. Photographing planetary nebulae and galaxies, however, is mostly the preserve of the cooled CCD (see page 21) on a medium- to large-aperture telescope; it must be either manually guided or controlled using an autoguiding system with a CCD to correct the telescope's drive rate.

Deep sky nomenclature

Most of the bright deep sky objects are included in Messier's catalog of 110 objects, which have M numbers. The majority of other deep sky objects observable with small- and medium-sized telescopes have either NGC or IC numbers. These refer to the New General Catalogue published by J. L. E. Dreyer in 1888, and to its Index Catalogue extension. Most of the Messier objects also have NGC numbers. While M numbers are assigned more or less at random throughout the sky, they do not appear south of declination –35°, which was the practical southern limit of Messier's observations. The NGC numbers are assigned in approximate order of right ascension, and include a wide range of objects, bright and faint, throughout the sky, so adjacent numbers may refer to objects widely separated in location and type.

The other major catalog is a recent compilation by British astronomer Sir Patrick Moore of 109 bright deep sky objects that are easily observable with small- and medium-sized telescopes. It is known as the Caldwell Catalogue. The Go To lists of Meade telescopes in particular use Caldwell numbers.

CHAPTER 8
THE CONSTELLATIONS

Magnitudes: 6 5 4 3 2 1 0 brighter than 0
Double or multiple stars
Variable stars
Open clusters
Globular clusters
Bright nebulae
Dark nebulae
Planetary nebulae
Galaxies

On the following pages are maps of the 50 most important constellations. Not every constellation is included because some have few objects of interest within them. The constellations are arranged into groups to make it easier to plan your observing, with those in the same part of the sky kept together.

In each case there is a range of objects to look for with small to medium telescopes, with illustrations for each one. Some of these illustrations were made with amateur instruments both large and small, while others are taken through some of the largest telescopes in the world, to give you some idea of what the faint misty object you are seeing looks like in depth. The descriptions are intended for observers under average modern conditions. Many observing guides tell you what you can see under ideal conditions, though few of us have such luxury. Even a tiny telescope will show faint objects under perfect skies, but most observers these days have to put up with some light pollution.

The descriptions of objects often refer to telescopes as being small, medium or large, and in this book the terms refer mostly to instruments used by amateurs. We have adopted a general classification in which small means telescopes from 50 to 100 mm aperture, medium means 110 to 200 mm, while large means any telescope bigger than 210 mm – usually up to about 450 mm in practical terms. The terms low, medium and high magnification are also used, but they are harder to define precisely as they depend on the instrument being used. For many catadioptric telescopes at f/15, the lowest power usually available is ×75–×100, which for a small refractor would be a medium to high power. But in general, low power means ×25–×75, medium means ×100–×200, and high means greater than ×200.

The tables of information about the objects described give a visual magnitude (M_V), magnification, and distance for each object. Each of these figures is only a guide. The magnitude and magnification figures are intended to give some idea of what you are looking for, the magnification quoted being a suggestion for getting a good view of the object, rather than the minimum that will show it. Distances are also unreliable – different sources often give widely differing values, so don't be surprised if you see other values quoted elsewhere. Establishing a good distance scale is one of the most challenging aspects of modern astronomy.

GROUP I

Ursa Major

Call it the Big Dipper or call it the Plough, the pattern of the seven brightest stars in the constellation of Ursa Major is the best known in the northern hemisphere. These stars are by no means the brightest in the sky, but they are visible all night from North America and Europe. They are also easy to recognize, and once you have spotted the familiar grouping you can draw a line through the two right-hand stars, known as the Pointers, to point roughly to the Pole Star. They are to the northern hemisphere what the Southern Cross is to the southern hemisphere.

▲ *The stars of the Big Dipper make an ideal guide to the Pole Star, as shown here. The line misses Polaris by a few degrees.*

The seven stars are just a small part of Ursa Major. The group was known as a bear from ancient times, though like the lesser bear, Ursa Minor, this one has an amazingly long tail. According to Greek legend the bear was originally a beautiful girl named Callisto, who was seduced by Zeus. For her sins she was turned into a bear and thrown into the heavens, and her tail was drawn out in the process.

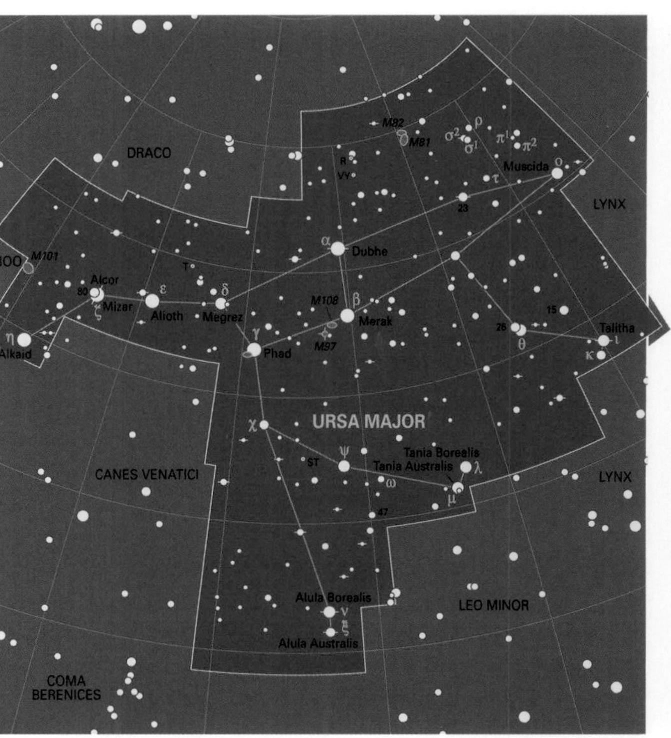

Object	Type	Mv	Magnification	Distance
Mizar	Double star	2.3	×50	86 light years
M81	Galaxy	6.9	×75	12 million light years
M82	Galaxy	8.4	×75	12 million light years
M101	Galaxy	7.9	×75	22 million light years
M97	Planetary nebula	9.9	×100	2000 light years

To the peasants of Britain centuries back, the tail of the bear resembled the handle of the wooden plow that nowadays is mostly seen on the walls of pubs of the same name, while to the pioneer Americans it was the handle of the dipper used to ladle water. Neither appliance is in common use today, but the names live on.

▲ *Though different sizes in a photograph, the galaxies M81 (left) and M82 are similar sizes when seen in light-polluted skies because the spiral arms of M81 are hard to see.*

Galaxies M81 and M82

This is a delightful pair of galaxies, not just because they are fairly bright but also because they have contrasting shapes that show up even using small telescopes. Users of Go To telescopes should have no trouble finding them, as they are large enough to be seen in a low-power field of view. To find them in binoculars, take a diagonal line between the stars Gamma and Alpha and you will spot them the same distance beyond Alpha, just short of a pair of stars.

▲ *A drawing of galaxy M101 made using a 450 mm Dobsonian with a power of ×102.*

M81 is a classic spiral galaxy, though with most telescopes all you see is the oval haze of the central regions. M82 by contrast is termed a peculiar galaxy. In a telescope it looks quite different from M81, as a short, stubby bar, and photos reveal signs of what looks like an explosion at its center. It is what is known as a starburst galaxy, and star formation is taking place there at a great rate. One theory is that this activity was triggered by an interaction with M81 some 100 million years ago.

Galaxy M101

Very few galaxies actually do show spiral arms when seen through a telescope. Quite often, as when viewing M51 (page 86), you imagine that you can see the spiral arms, but that is only because you know they are there. But M101 is one of the small band of galaxies that are indeed large and bright enough to reveal their arms in amateur telescopes.

The drawing was made using a large reflector at a dark site, but telescopes of 300 mm or larger should show the spiral arms under the right conditions. In a dark sky, M101 seems difficult to miss even when viewed with binoculars, but because of its fairly low surface brightness it can be very hard to find under light-polluted conditions, even with a Go To telescope.

Double star Zeta Ursae Majoris (Mizar)

Mizar is the middle star in the handle of the Dipper or Plough. With the naked eye you can see a 4th-magnitude star nearby, Alcor. Through a small telescope, you can see that Mizar is a true double star in its own right.

▲► *The Big Dipper or Plough (above). The inset (right) shows the telescopic view of Mizar and Alcor. The two stars of Mizar are separated by 14 arc seconds.*

M97, Owl Nebula

Though this is a well-known planetary nebula, many amateur astronomers have never actually seen it because it is particularly small and faint. The photograph shows why it got its name: two dark areas within the disk of the planetary look like the eyes of an owl. You need at least a 250 mm telescope and no haze to be able to see these with your own eyes.

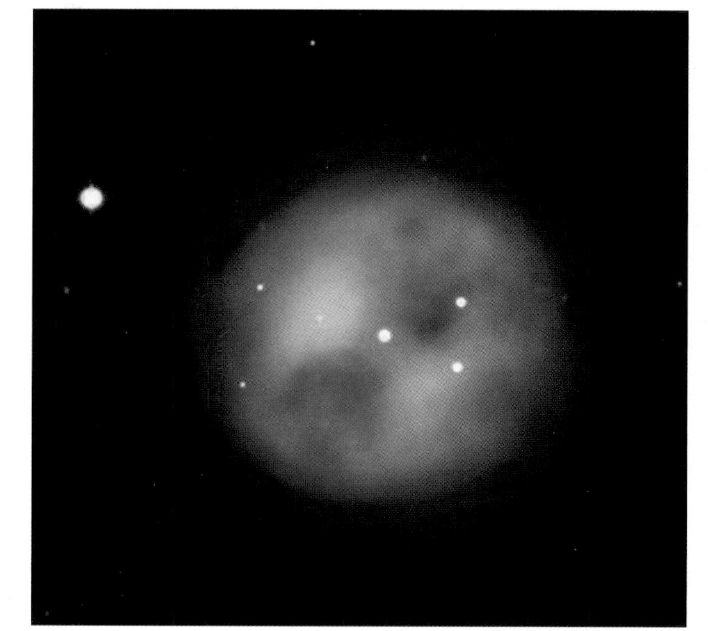

▲ *A photograph with a 0.9 m telescope shows the green color of M97 and the two voids that resemble the eyes of an owl.*

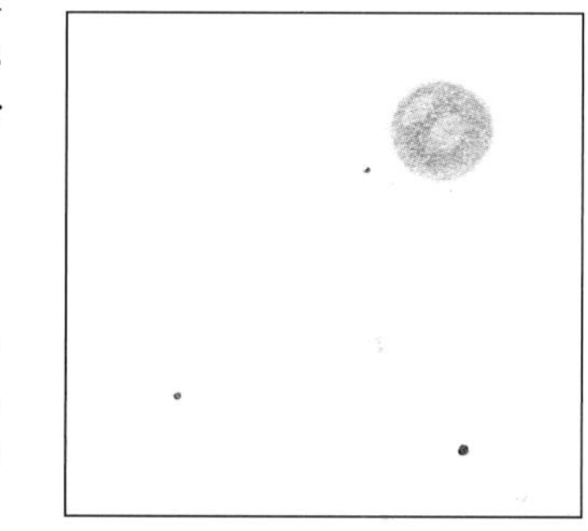

◄ *Seen in a 450 mm reflector in good skies, the "eyes" are visible, but they have been seen even with 250 mm telescopes.*

Canes Venatici

These hunting dogs were placed in the sky comparatively recently, in 1687 by Johannes Hevelius, and they are there to help Boötes, the herdsman, to the east. The brightest star in Canes Venatici has a curious history. Alpha is also known as Cor Caroli, meaning Charles' Heart. There are two stories about how it got this name. One says that it was named after the English King Charles II, who was restored to power on May 29, 1660, when the star was said to be particularly brilliant. But another story holds that it commemorates his predecessor, the unfortunate Charles I, who was beheaded in 1649. The star is indeed slightly variable, so who knows what may happen if it chooses to flare up again! Cor Caroli is also an easy double star.

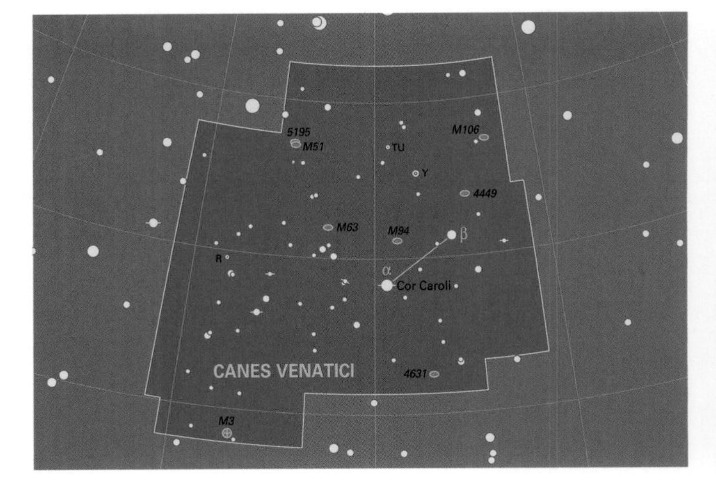

Globular cluster M3

A bright globular cluster, M3 is one of the best in the northern sky. It is said to be the original object that Charles Messier mistook for a comet. But when he realized that he had been fooled, he set about drawing up the catalog of misty objects that now bears his name. You will need about a 100 mm telescope to turn the circular haze into actual stars.

▲ *When photographed using a large telescope, M3 shows a burnt-out center. Compare this photograph with the drawing on page 76, which was made with a 150 mm reflector.*

Galaxy M94

This is a spiral galaxy, with tightly wound arms. Even a small telescope, however, gives a hint of its spiral nature. It is easy to find even without a Go To telescope because it forms a triangle with the two brightest stars in Canes Venatici.

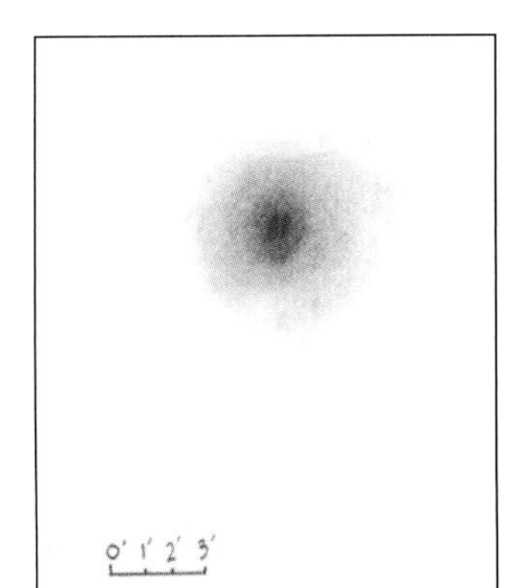

◀ *Galaxy M94 as seen through a 76 mm refractor.*

Galaxy M51

One of the most famous galaxies in the sky, M51 is also known as the Whirlpool Galaxy. It is the object that was first seen to be in the shape of a spiral, by Lord Rosse in 1845, though it was not known to be a galaxy at the time.

Lord Rosse used a giant 72-inch (1.8 m) telescope, but today amateurs with telescopes as small as 300 mm can repeat the observation. This is simply because today's telescopes, with mirrors made of glass and coated with aluminum, are much more reflective than the polished metal mirror used by Rosse. Even a 75 mm telescope will show the companion galaxy, NGC 5195.

M51 is most easily found by star-hopping from the end star of the handle of the Big Dipper or Plough.

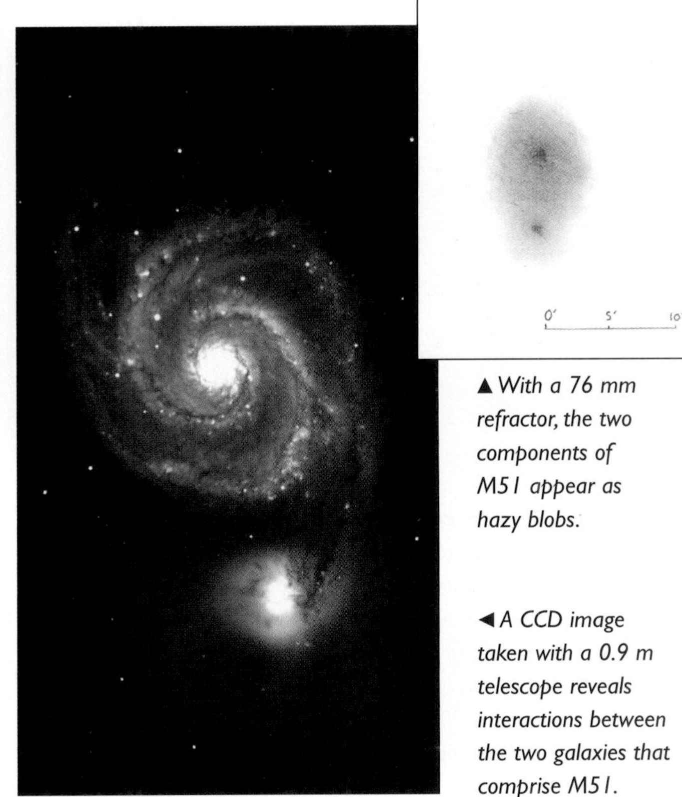

▲ *With a 76 mm refractor, the two components of M51 appear as hazy blobs.*

◀ *A CCD image taken with a 0.9 m telescope reveals interactions between the two galaxies that comprise M51.*

Object	Type	Mv	Magnification	Distance
M3	Globular cluster	3.2	×25	33,900 light years
M51	Galaxy	8.4	×50	26 million light years
M94	Galaxy	8.2	×75	14 million light years

Draco

Dragons are meant to be fearsome, but the celestial version is rather faint, if large. It winds across a swathe of the northern sky around Ursa Minor, between the two bears. Its most prominent feature is its head, which is represented by four stars above Hercules, the hero who slew the dragon in mythology. Another star in Hercules makes a quite distinctive diamond shape with Draco's three brightest stars. The brightest star in Draco is actually Gamma, in the head of the dragon, and the lettering of the stars in Draco is quite chaotic, following neither their order along the dragon nor the normal bright to faint sequence. This is probably the celestial dragon's most fearsome aspect.

Double star Nu Draconis

The faintest star of the Head of Draco, Nu Draconis is a pretty double star seen with binoculars. Two virtually identical white stars look like car headlights separated by just over a minute of arc. If you need to know what 1 arc minute looks like, this star shows you.

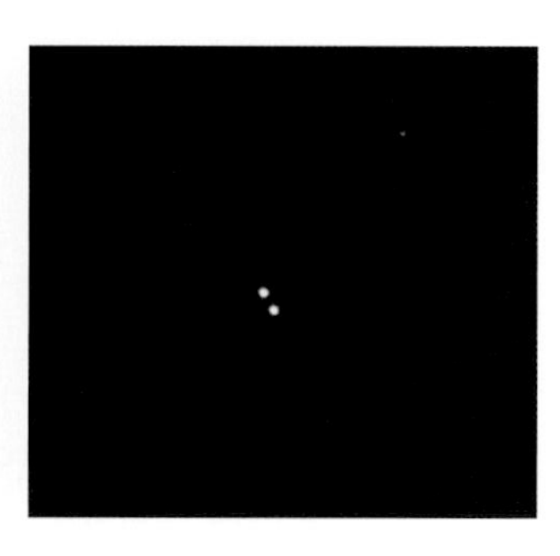

◄ *The wide double star Nu Draconis is a pretty sight even with low-powered binoculars.*

Planetary nebula NGC 6543

This was just another planetary nebula, though a fairly bright one, until the Hubble Space Telescope photographed it and revealed it as one of the most distinctive objects in the sky – the Cat's Eye Nebula. Through a medium-sized telescope it is a bluish disk, almost as bright as but much smaller than the Ring Nebula in Lyra. The first Hubble versions showed it as a red eye, but in a new photo published in 2004 it is blue, with numerous shells surrounding the star.

Object	Type	Mv	Magnification	Distance
Nu Dra	Double star	1.2	×10	108 light years
NGC 6543	Planetary nebula	8.1	×250	3600 light years

►▼ *The Cat's Eye Nebula, NGC 6543, photographed by the Hubble Space Telescope (right). Even with an amateur telescope (below), in this case a 450 mm, the cat's eye shape is evident.*

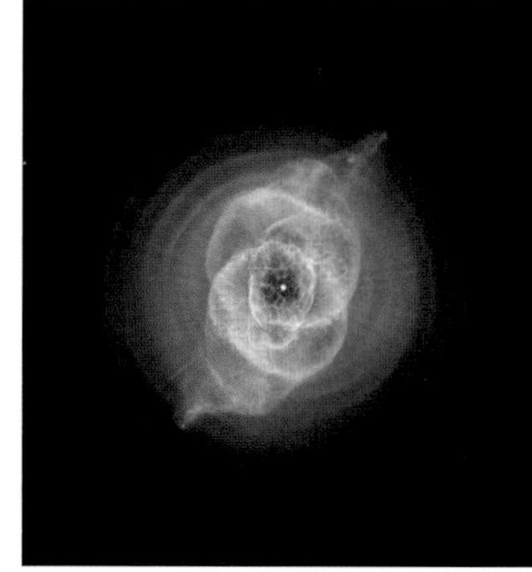

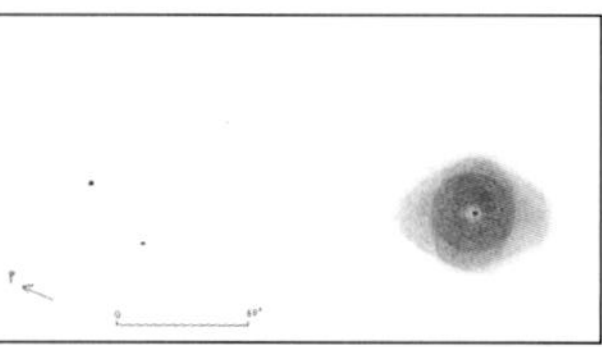

Ursa Minor

Ursa Minor owes its fame entirely to its brightest member – the Pole Star, Polaris. Though this is not a particularly bright star (contrary to popular myth), its location close to the sky's north pole has made it the most significant in the sky for hundreds of years. Since late Roman times it has been the Pole Star, though only since medieval times has it been particularly close. To mariners in particular, a glimpse of Ursa Minor and the Pole Star has told them the direction of north and their latitude. Even today, many telescope mounts have small telescopes built in that allow you to use Polaris to align your equatorial mount correctly. But beware of Kochab, the star at the other end of Ursa Minor, which is almost equal in brightness to Polaris. Choose it by mistake and you will be way off.

The rest of Ursa Minor is a fairly faint and undistinguished constellation that roughly resembles its more illustrious neighbor, Ursa Major, with a rectangle and a tail of stars that curves off to end in Polaris itself. In North America it is also known as the Little Dipper, though no one in Britain calls it the Little Plough. Because Ursa Minor is always at more or less the same altitude in the sky from any particular location in the northern hemisphere, many people use it as a guide to the transparency of the sky. You can look for its stars at any time of night or year and judge how much they have been affected by the absorption of the atmosphere.

Alpha Ursae Minoris (Polaris)

Polaris is probably observed more than any other star in the sky because so many people use it to align their mountings. Few

Object	Type	Mv	Magnification	Distance
Polaris	Star	2.0	Naked eye	431 light years

pay it any attention once it has served its purpose, but it has its merits. Nearby lies a 6th-magnitude star, a member of an attractive circlet of stars sometimes known as the "Engagement Ring," which has Polaris itself as the diamond. Polaris is a Cepheid-type variable (see page 75), but its brightness changes are very slight.

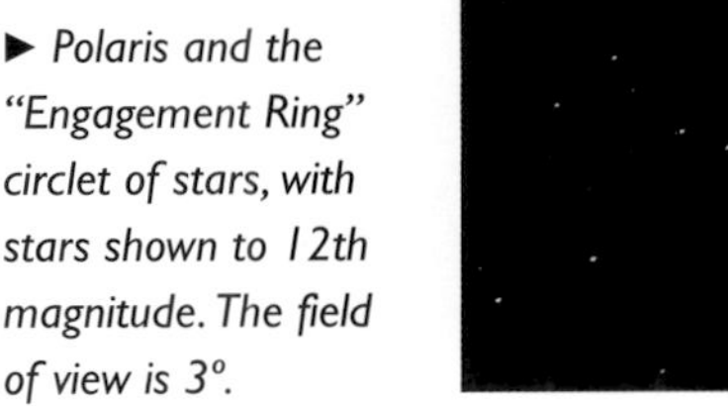

► *Polaris and the "Engagement Ring" circlet of stars, with stars shown to 12th magnitude. The field of view is 3°.*

Cepheus

Most northern-hemisphere constellations, such as Cassiopeia and Cygnus, have clear patterns that make them easy to spot. But Cepheus is more like a southern constellation – a mass of stars of different brightnesses with no distinguishing features. Despite including a chunk of the Milky Way, it has no bright deep sky objects. It is better known for its stars, one of which has made the constellation a household name, at least among astronomers. Cepheus himself was a king in the Perseus legends, the husband of Cassiopeia.

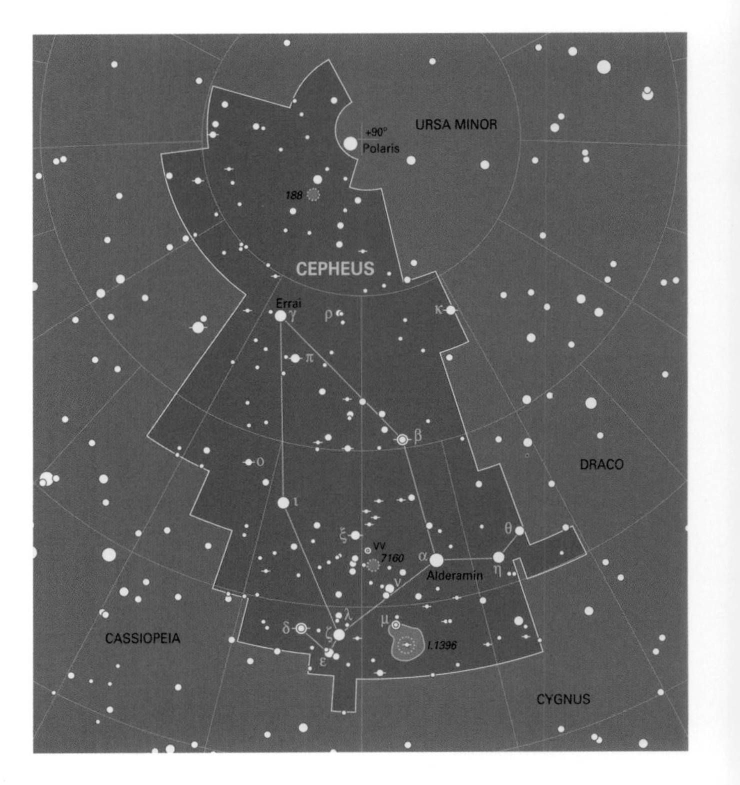

Variable star Delta Cephei

This star, at the apex of a little triangle at the southern end of the constellation, was one of the first variable stars to be recognized. It changes in brightness between magnitudes 3.5 and 4.4 over a regular period of 5 days 9 hours, with a slow decline and a more rapid increase. This is typical of a particular class of variable stars known as Cepheid variables, after this star.

Cepheid variables are particularly useful to astronomers because the brighter they are in real terms, the longer they take to vary. This means that any variable star showing this characteristic pattern of variation is advertising its true brightness. By comparing its true brightness with the brightness it appears in the sky, it is an easy matter to work out its distance. One of the tasks of the Hubble Space Telescope was to detect Cepheid variable stars in the Virgo Cluster of galaxies, thus establishing its distance.

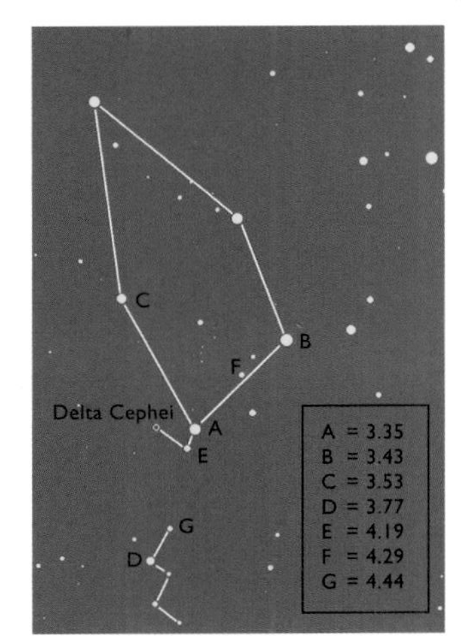

► *You can follow the brightness variations of Delta by comparing its brightness with those of nearby stars. A light curve of the star is shown on page 75.*

Mu Cephei, Garnet Star

Star colors are usually rather elusive. Diagrams in books may show red giant stars as the color of traffic lights, but in fact most are actually no redder than a 100-watt light bulb. Mu Cephei is known as the reddest star visible to the naked eye in the northern hemisphere, and in binoculars or a telescope its color is unmistakable. But it is not bright enough to appear particularly red to the naked eye. William Herschel famously described it "a very fine deep garnet color," so it is often called "the Garnet Star."

Nebula IC 1396

A large and impressive nebula, IC 1396 is virtually impossible to see but quite easy to photograph. It starts to show up on photographs made using a standard lens and a few minutes' exposure at 400 ISO, with Mu Cephei at one edge. The distances of both Mu and IC 1396 are uncertain, so it may not be connected with the nebula.

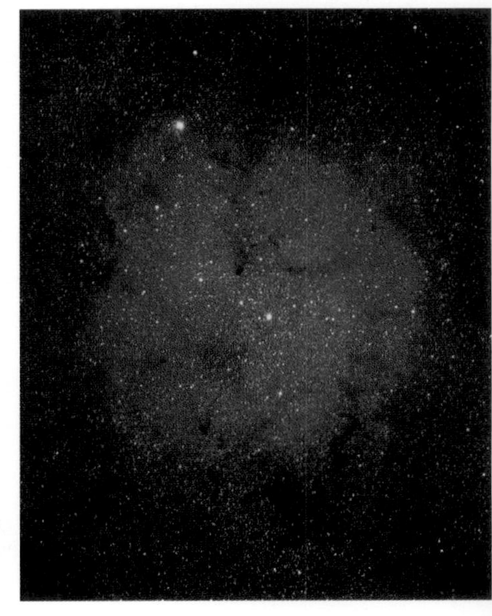

► *Both Mu Cephei and IC 1396 feature in this view. Mu is the red star at top left. The time exposure also reveals the diffuse nebula IC 1396, which is over 2° across.*

Object	Type	Mv	Magnification	Distance
Delta Cep	Variable star	3.5–4.4	Naked eye	863 light years
Mu Cep	Star	4.2	Naked eye	5260 light years
IC 1396	Nebula	–	Photographic	2450 light years

GROUP 2

Hercules

This is a large, faint, and sprawling constellation in the mid northern sky. Its most recognizable feature is a pattern of four stars known as the Keystone, which lies midway between the more easily spotted semicirclet of Corona Borealis and the bright star Vega. Once you have found the Keystone, look for the extensions southward to Beta Herculis (Kornephoros) and Alpha Herculis (Rasalgethi). The Keystone is also your signpost to the two globular clusters, M13 and M92, which are the main attractions of the constellation.

Hercules is an ancient constellation, featuring in many of the Greek legends about the heavens. But the poor chap is depicted upside down in the sky, this being the only way he can rest his foot on the head of the dragon, Draco, that he slew.

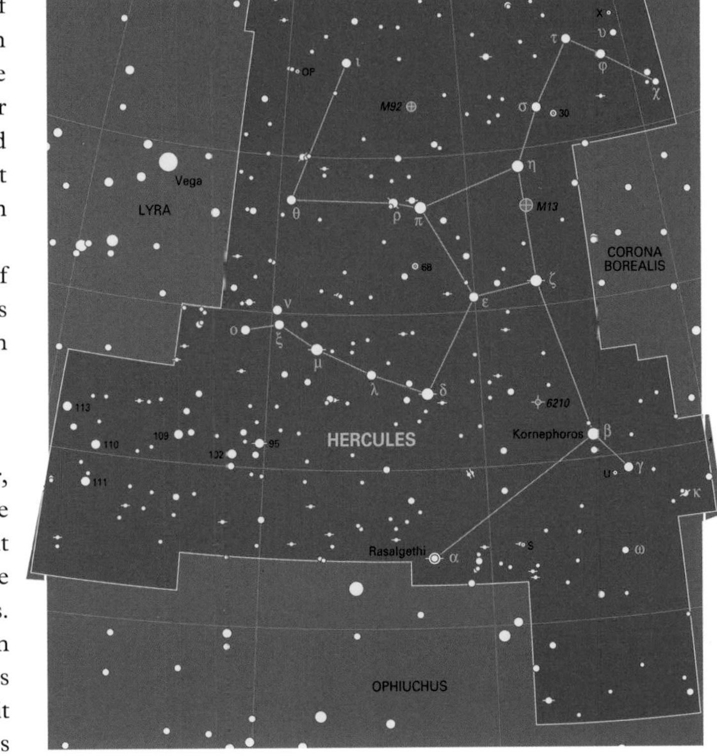

Globular cluster M13

Sometimes called "the Great Hercules Cluster" or similar, M13 is the brightest globular cluster in the northern half of the sky (though southern observers have been heard to sneer at it in comparison with the great southern globulars). It is a fine sight in a medium-sized telescope, appearing as a ball of stars. Find it by looking about two-thirds of the way up the western edge of the Keystone. In small instruments and binoculars it is a hazy circular blur, but the larger your telescope, the more it turns into individual stars. Photos made with large telescopes usually make the center appear white as all the star images run together, but through even a large telescope you see only individual stars as pinpoints of light. The cluster has around a million stars in all.

◄ *A close-up of M13 taken with a CCD imager on a 130 mm refractor in Dengie Marshes, Essex, UK.*

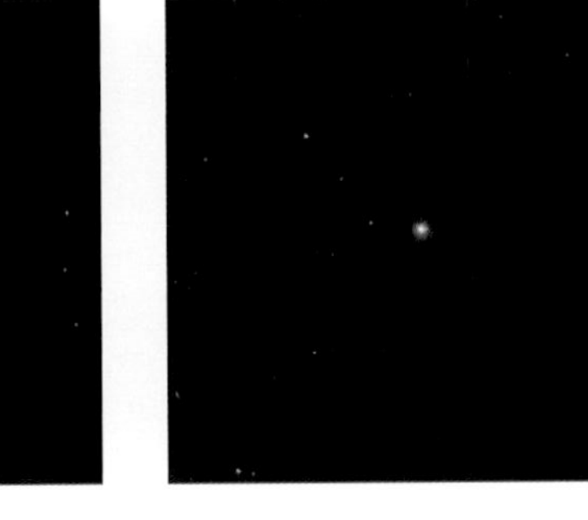

▼ *M13 (left) and M92 offer an interesting comparison. These telephoto lens shots give the same impression as you would get visually with a small telescope.*

Globular cluster M92

Though less rich than M13, M92 is still an enjoyable sight. It is easy to spot in binoculars in a good sky, though city observers will struggle as it may be difficult to distinguish M92 from a star, being quite a compact object. It makes a triangle with the two northernmost stars of the Keystone. It is easy to compare M92 with M13 as they are only 9° apart. M92 is renowned for containing some of the most ancient stars of any globular cluster in our Galaxy. At around 14 billion years old, they are similar in age to the Universe as a whole. See also the image on page 77.

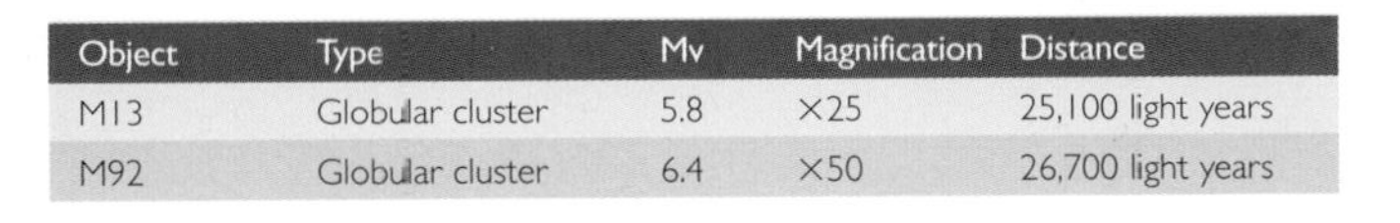

Object	Type	Mv	Magnification	Distance
M13	Globular cluster	5.8	×25	25,100 light years
M92	Globular cluster	6.4	×50	26,700 light years

Ophiuchus

Ask a stargazer to point to Ophiuchus and they will probably wave a hand at a barren area of sky somewhere above Scorpius and say "It's around there." Though its brightest stars are of 2nd magnitude, similar to those of, say, Cassiopeia, they are spread over such a large area of sky that there is no easily recognized pattern other than a misshapen pentagon. Possibly its most recognizable feature is a pair of stars, about 20° above the bright stars of Scorpius, named Yed Prior and Yed Posterior, meaning "Hand-in-front" and "Hand-behind" in a mix of Arabic and Latin. Though these stars are unrelated to each other, their proximity in the sky makes an eye-catching pair that marks the western end of Ophiuchus. The constellation represents Aesculapius, a mythical healer, whose snake-entwined staff is still a symbol for the medical profession.

One peculiarity about Ophiuchus is that it is nowadays a major constellation of the zodiac – the band across the sky along which you find the Sun, Moon, and planets. The Sun is actually within the borders of Ophiuchus between November 30 and December 18, whereas it spends only the preceding six days in Scorpius. The dates given in the "Star Signs" that you read in the papers have nothing to do with the actual positions of the heavenly bodies in the sky, and the definitions of the boundaries of the constellations have changed over the years. Ophiuchus extends both above and below the celestial equator, so it is visible from both hemispheres.

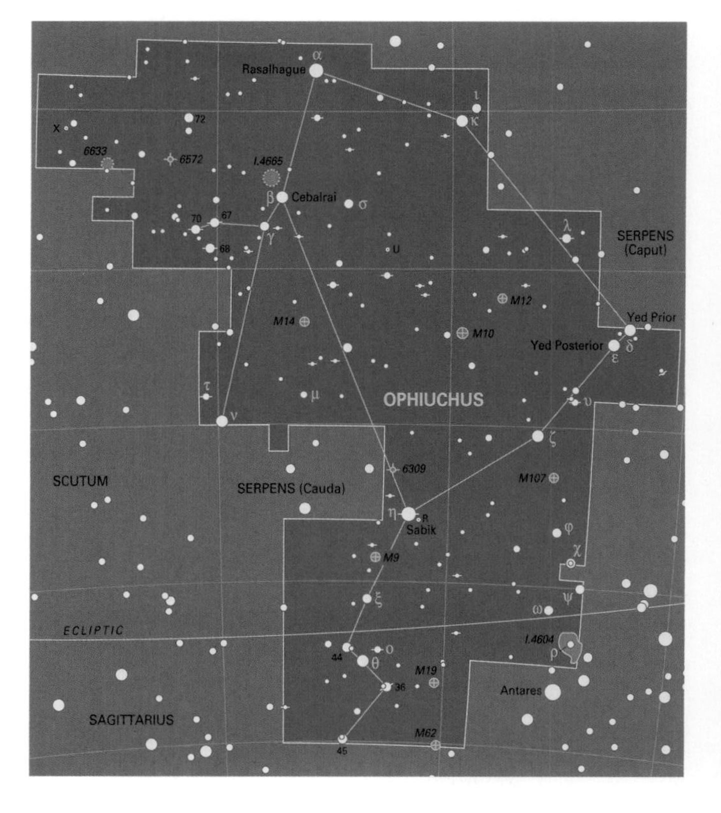

Open cluster IC 4665

▲ *The area of open cluster IC 4665 in Ophiuchus, showing stars down to magnitude 10.*

Most objects in the Index Catalogue (the successor to the New General Catalogue or NGC) are rather faint objects. IC 4665 is an exception: a scattering of a dozen or so stars of around 7th magnitude, it is easily visible in binoculars. But the stars are so loosely clustered that in a telescope you may miss it.

Milky Way around Rho Ophiuchi

The Milky Way passes through Ophiuchus, and the area just above Antares in Scorpius is a favorite among photographers when there are dark skies, for in this area are beautiful bright and dark nebulae of differing colors. But in average skies little is visible. In the photo, the bright area at the bottom is the star Antares in Scorpius, while to its right is the globular cluster M4. Rho Ophiuchi is at the top, surrounded by a blue nebula.

▲ *Nebulosity around Antares and Rho Ophiuchi is a favorite target for astrophotographers fortunate to have good dark skies.*

Globular clusters M10 and M12

These two clusters are interesting to compare because they are of similar brightness and are within the same field of view of many binoculars, which will show them looking like hazy stars in fairly dark skies. They are the brightest of many globular clusters in Ophiuchus. M10 is more condensed, while M12 appears larger. As with all globulars, the larger the telescope you use to view them, the better they appear.

► *Binoculars will show the globular clusters M10 (lower left) and M12 (upper right) within the same field of view. This image shows stars to magnitude 13.5, which is fainter than the normal binocular limit.*

Object	Type	Mv	Magnification	Distance
Rho Oph complex	Nebula	–	Photographic	400 light years
M10	Globular cluster	6.6	×25	14,300 light years
M12	Globular cluster	6.7	×25	16,000 light years
IC 4665	Open cluster	4.2	×10	1400 light years

Scutum

A distinctive C-shape of stars at the south end of Aquila directs you to Scutum, though most of the C-shape is actually in Aquila itself. The rest of this small constellation has no particular shape. However, Scutum contains a star cloud that is the brightest part of the Milky Way seen from northerly latitudes (from which the brighter star clouds of Sagittarius are too low in the sky to be prominent). Being just south of the celestial equator Scutum's star cloud is equally accessible from either hemisphere.

M11, Wild Duck Cluster

This beautiful cluster, visible in binoculars as a small misty patch, needs a telescope to do it justice as most of its stars are of magnitude 10 or 11. Its name comes from the V-shape of its brighter stars, resembling wildfowl in flight. Only a few hundred of its nearly 3000 stars are within reach of amateur telescopes.

Object	Type	Mv	Magnification	Distance
R Sct	Variable star	4.2–8.6	Naked eye	3000 light years
M11	Open cluster	6.3	×50	6000 light years

◀ *The Wild Duck Cluster, M11. The V-shaped pattern of stars is best seen using a telescope.*

Variable star R Scuti

This is a favorite among variable star observers, because its location near the Wild Duck Cluster in a quadrilateral of stars makes it easy to find and to make estimates of its magnitude. It is the brightest of a type known as RV Tauri stars, and it spends much of its time at around magnitude 5. It may dim to magnitude 8 every four or five months, making the quadrilateral of stars look different, though for years on end its variations may be only a few tenths of a magnitude.

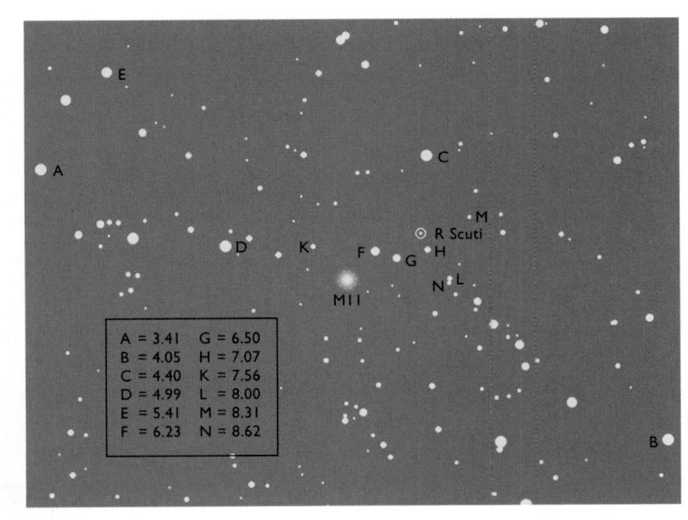

▲ *This chart of comparison stars for R Scuti is adapted from the one used by the British Astronomical Association's Variable Star Section.*

▲ *The Eagle Nebula, M16, photographed with the 4 m Mayall telescope at Kitt Peak, Arizona. The central dark features resemble an eagle, though some observers see them as a "Star Queen."*

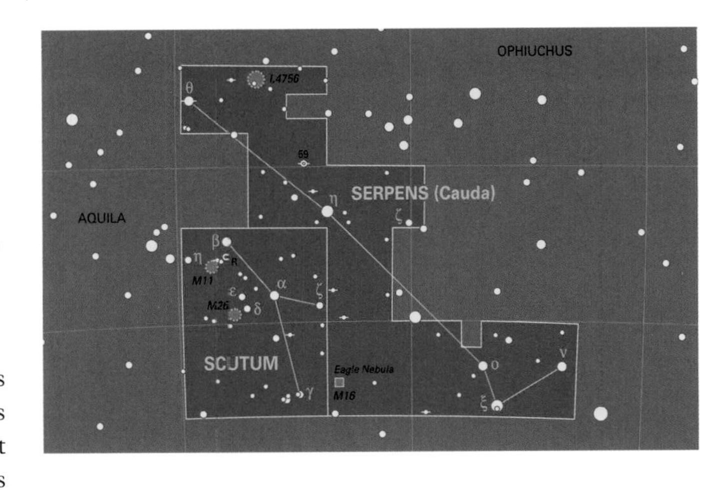

Serpens Cauda

The stars of Serpens Cauda – the Serpent's Tail – are faint and undistinguished. They form a line to the west of Scutum, but one of the stars apparently in the line is actually in Ophiuchus. The brighter head of Serpens is on the other side of Ophiuchus. The constellation has little to offer the visual observer, as it is almost totally in the direction of a nearby dark cloud, part of the Great Rift (see page 13). Only at the extreme southern and northern ends are there many stars visible in binoculars. The cluster IC 4756 is an attractive sight with a low-power telescope.

M16, the Eagle Nebula

This is the location of one of the world's most famous astronomical images, the amazing "Pillars of Creation" from the Hubble Space Telescope. But visually it is rather a disappointment unless you have dark skies and the right instrument. In average skies it appears as a run-of-the-mill open cluster, with perhaps a hint of nebulosity. Binoculars show it clearly, though the individual stars are not obvious. Wide-field photographs pick up the red glow typical of a starbirth region, but close-up deep exposures show the darker lanes in the nebulosity to the south of the cluster. These "elephant trunks," as they are generally called, are pillars of denser gas within which stars are forming.

The Eagle Nebula gets its name from a fancied resemblance to an eagle with raised head and outstretched wing. These features can be glimpsed in large amateur telescopes but they are best enjoyed on photographs.

▶ *The stunning "Pillars of Creation" at the center of the Eagle Nebula, M16, photographed by the Hubble Space Telescope. The left-hand trunk is the "head of the eagle."*

Object	Type	Mv	Magnification	Distance
M16	Open cluster/Nebula	6.4	×75	7000 light years

Sagittarius

A jewel among constellations, Sagittarius includes the richest areas of the Milky Way and contains more bright deep sky objects than any other region of the sky. But to many northern observers, Sagittarius is hard to observe. From much of North America and Europe, it never rises more than 10° or 20° above the horizon. As a result, many of its wonders are lost in haze and are dismissed in favor of lesser objects higher in the sky.

The stars of Sagittarius are not particularly bright – the brightest, Kaus Australis, is only about 37th in the league table of bright stars. To many people, Sagittarius is most easily spotted by what is called the Teapot, a pattern of eight stars that really does resemble a teapot. Most of the numerous deep sky objects in the constellation are near this area, within the plane of the Milky Way. In a dark sky, with binoculars or a low-power telescope, the whole area is a superb sight with countless stars.

Sagittarius represents a centaur, not to be confused with Centaurus itself. He is an archer, aiming his arrow at the heart of neighboring Scorpius – the spout of the teapot is actually the point of the arrow.

The ecliptic runs through Sagittarius, this being its most southerly point.

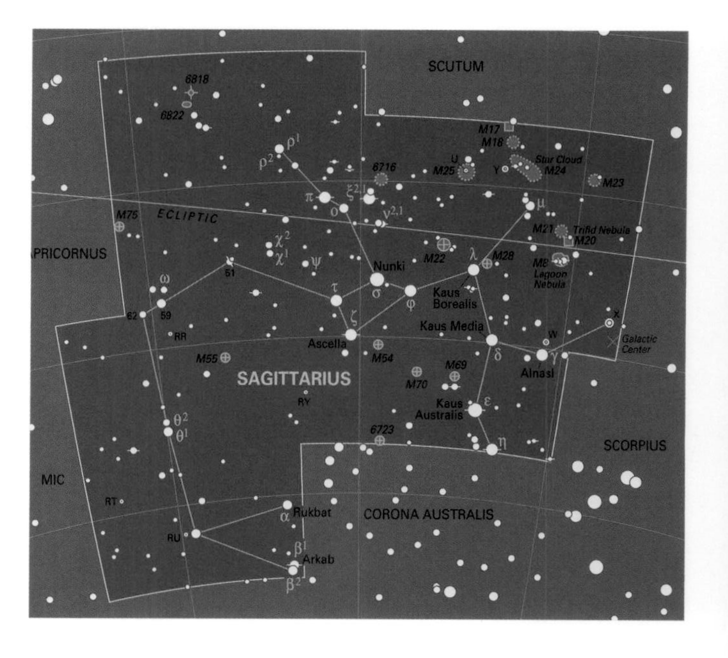

M8, the Lagoon Nebula

This is one of the brightest nebulae in the sky, following only the Eta Carinae Nebula and the Orion Nebula. In a good sky it is easily visible with the naked eye. In binoculars M8 looks like a cluster surrounded by an oval haze. Telescopically the dark central channel – romantically seen as a lagoon by 19th-century popularizer Agnes M. Clerke – is easy to see with low powers. The nebula is larger than the apparent size of the Full Moon under good conditions.

The Lagoon Nebula lies at the southern end of a ragged line of nebulae and clusters, which includes most of the other diffuse nebulae in Messier's catalog. It makes an excellent starting point for a tour of deep sky wonders. In Chapter 1, on page 5, you can compare a photograph of M8 taken with a large amateur telescope with a drawing made using a 300 mm telescope.

M20, the Trifid Nebula

This nebula is more difficult to see than the Lagoon, though it lies within the same binocular field of view. It is smaller and fainter than M8, and requires medium apertures for good views. The Trifid Nebula gets its name from its three dark lanes; the word is Latin for "cleft in three," reminiscent of John Wyndham's classic sci-fi story *The Day of the Triffids*, about three-footed man-eating mobile plants – though there is no connection. The northern part of M20 is a blue reflection nebula. Though both are rather disappointing visually, they are easy to photograph, and the contrasting colors jump out at you.

► *Another famous poster object, visible within a short distance of both M16 and M8. M20 is one of the brightest examples of a reflection nebula in the sky.*

▲ *From left to right, the open cluster M23, the star cloud M24 and the open cluster M25. Some books refer to a smaller cluster, NGC 6603, at the heart of the star cloud as being M24. At the top of the picture is the Swan or Omega Nebula, M17.*

Open clusters M23, M24, and M25

Lying across the north–south line of diffuse nebulae is this band of three clusters. M24, the central one, is sometimes known as the Little Sagittarius star cloud. It is really a large bright patch of the Milky Way, though there are a number of stars embedded in it which appear as a cluster in a small telescope. All three clusters are easily visible with binoculars and are best viewed at low power in a telescope.

Object	Type	Mv	Magnification	Distance
M8	Nebula	4.6	×25	5200 light years
M17	Nebula	6.0	×25	5000 light years
M20	Nebula	9.0	×25	5200 light years
M22	Globular cluster	5.1	×25	10,400 light years
M23	Open cluster	6.9	×25	2150 light years
M24	Open cluster	4.6	×10	10,000 light years
M25	Open cluster	6.5	×25	2000 light years

M17, the Swan or Omega Nebula

This nebula goes by several aliases. To our classically educated forebears it was a Greek capital omega, while to others it is the number 2, or a horseshoe. But many people prefer to see it as a swan. It is bright and easy to spot in binoculars, while a low power on a 114 mm telescope readily reveals the swan shape. To its south lies a rather sparse cluster, M18, while to its north is M16 in Serpens.

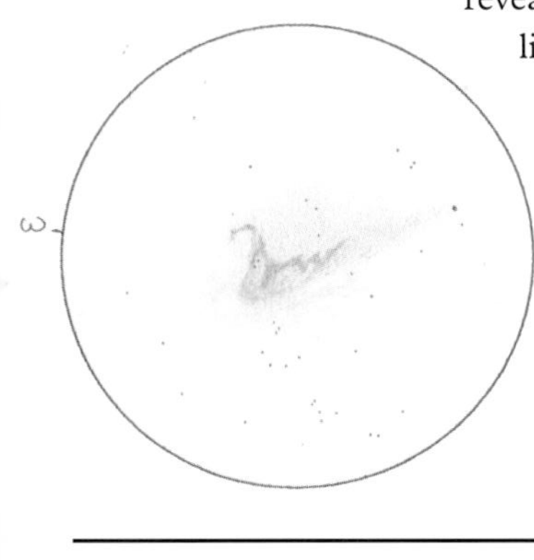

◀ *One of the brightest diffuse nebulae, the Swan or Omega Nebula, M17, has a characteristic shape. This drawing was made using a 300 mm reflector.*

Globular cluster M22

This is the brightest and largest of the globular clusters visible from the northern United States and northern Europe, though M13 in Hercules is often given that title. Being quite far south, however, M22 suffers, for many observers, from being low in the sky, which makes it appear less spectacular. M22 is the brightest of several globular clusters that can be found in Sagittarius.

▲ *If it were not for its southerly declination, globular cluster M22 would be better known among northern observers.*

GROUP 3

Cygnus

This celestial swan flies forever down the Milky Way, its long neck and wings outstretched. It is also referred to as the Northern Cross, though it is much larger than the better-known Southern Cross. The bright star Deneb marks its tail. This is a remarkable star, though telescopically it looks no different from any other. Deneb is a true searchlight, and its light dominates this part of the Galaxy. Though its distance is rather uncertain, a recent estimate puts it at over 1400 light years away. Compare that with Altair, a bright star some way south of Cygnus. Altair is a little brighter in the sky but is only 17 light years away – about 80 times closer.

In addition to containing the most distant bright star, Cygnus also contains one of the closest stars, 5th-magnitude 61 Cygni, just 11.4 light years away. This star was the first to have its distance measured. It is an easy double star, with components separated by 27 arc seconds, making it just visible as double with 10-power binoculars.

There are several interesting variable stars in Cygnus. Chi Cygni is a bright Mira-type star (see page 100). At its brightest, every 400 days or so, it is magnitude 3.3; when faint, it drops to magnitude 14.5 and is effectively invisible to all but quite powerful telescopes. Another now-you-see-it-now-you-don't star is W Cygni, which varies between magnitudes 5.0 and 7.6 in about 130 days. P Cygni is usually 5th magnitude, but it is massive and erratic, and has been known to brighten to magnitude 3.

Cygnus is a happy hunting ground for most types of deep sky object. For many people, just sweeping the area with binoculars is very enjoyable, as the Milky Way in this region is very rich. One feature of Cygnus is that its cross-shape makes it very easy to locate objects by star-hopping: there are easy routes to most objects within the constellation.

Object	Type	Mv	Magnification	Distance
Albireo	Double star	3.1	×10	420 light years
M39	Open cluster	4.6	×7	825 light years
NGC6826	Planetary nebula	8.8	×250	2200 light years
NGC7000	Nebula	–	×7	1600 light years
Veil Nebula	Supernova remnant	5?	×7	2600 light years

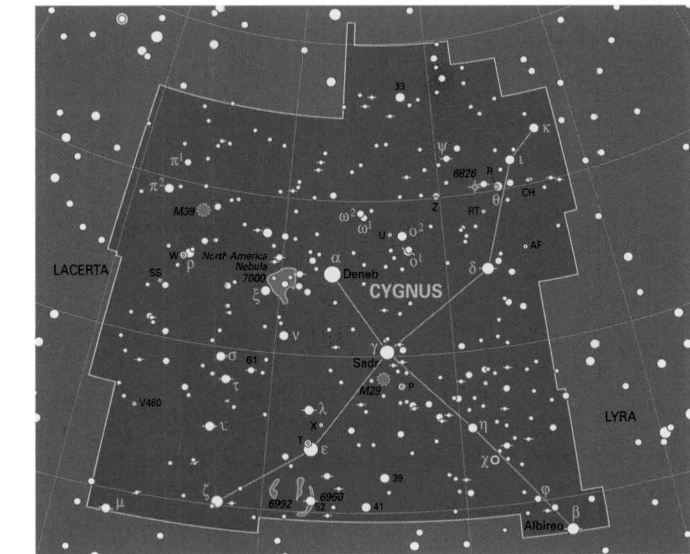

Double star Beta Cygni (Albireo)

Arguably the most popular double star in the sky, Albireo marks the head of the swan. It consists of two well-separated stars of which the brighter is yellowish and the fainter is bluish. This is no contrast effect (photographs show these colors clearly), though the proximity of the two makes the difference in color easier to see. The stars themselves are types K and B.

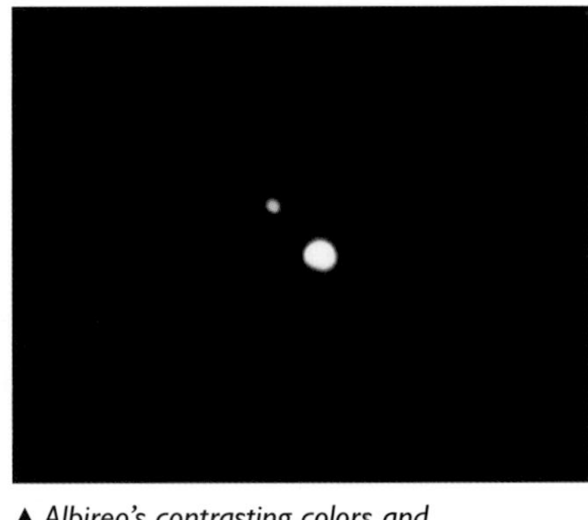

▲ *Albireo's contrasting colors and wide separation make it one of the most easily observed and attractive double stars in the sky.*

Open cluster M39

Not a spectacular open cluster, M39 is easy to find by following a line of stars northeast of Deneb. It is best seen with a low power.

◄ *Not far north of Deneb, M39 is in a rich part of the Milky Way.*

▼ *The Blinking Planetary is one of the smaller bright planetaries. It requires a very high power for the best views. This drawing was made using a 450 mm reflector with a magnification of 508.*

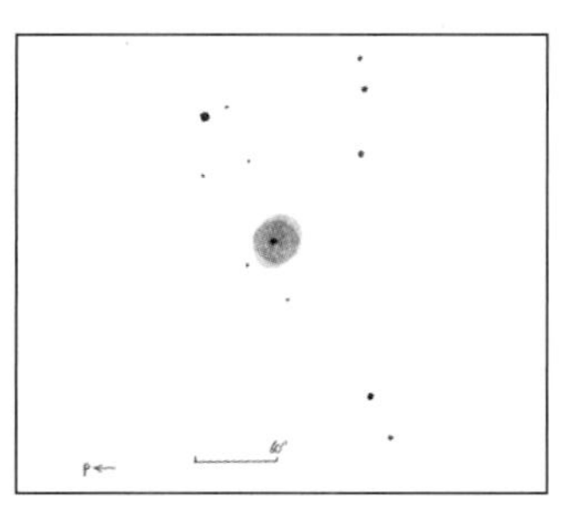

Planetary nebula NGC 6826, the Blinking Planetary

At 8th magnitude, this planetary is easy to find in binoculars or a finder at the end of a line of stars, though you may think it is just another star, as it is compact and round. It is about the same brightness as the more famous Ring Nebula in Lyra, but only half the diameter. Use considerable magnification – around 150 or more – and you will see that there is a central star within a circular bluish shell. This star is one of the brightest central stars of any planetary nebula, at about magnitude 11, and is visible with even a fairly small telescope. But as you stare at the star, you may notice an odd thing – the nebula surrounding it disappears! This is because when you look directly at the star it is bright enough to be seen using the cones in the center of your vision (see page 14), whereas with small telescopes the planetary surrounding it requires averted vision. If you look away, the nebula will reappear. This blinking effect depends on the aperture and the magnification.

NGC 7000, the North America Nebula

Some people say that they can easily see this nebula with the naked eye, but all that most can see is perhaps a slightly brighter area to the east of Deneb. Binoculars and an LPR filter might improve matters somewhat, but by far the easiest way to detect this impressive hydrogen gas cloud is to take a photograph of the area on fast film, with the camera's standard lens. As long as there is not excessive light pollution, an exposure time of a minute or so will reveal a pink glow with the unmistakable shape of North America. To its west is a fainter area known from its shape as the Pelican Nebula. In good conditions, the same setup may well show nebulosity around Gamma Cygni.

▲ *The North America Nebula is a favorite photographic target. Some types of film show it better than others, and digital cameras may not have enough red sensitivity.*

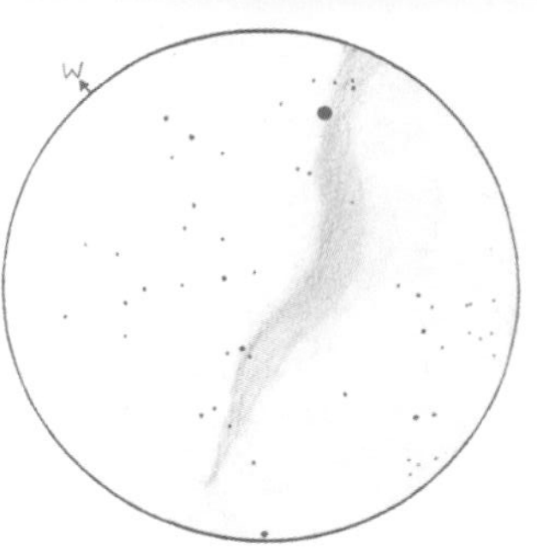

► *An O III filter was used on a 222 mm reflector for this drawing of part of the Veil Nebula. The star at the top is 52 Cygni. See also page 79.*

NGCs 6960 and 6992, Veil Nebula

This supernova remnant is a favorite target for deep sky observers and photographers. It is not difficult to see in country skies, but light pollution soon wipes it out. The usual method of finding it is to locate the star 52 Cygni, which lies right in the middle of the western arc of the nebula, NGC 6990. A light-pollution filter will help. Again, the object can easily be photographed, though because it is thinner than the North America Nebula, it is worth using a modest telephoto lens (about 135 mm or longer).

Lyra

Vega, to the west of Deneb, is one of the brightest stars in the sky, at magnitude 0. With Deneb and Altair it forms what is known to northern-hemisphere observers as the Summer Triangle, a large asterism (star pattern) that actually persists well into the early winter and is useful for helping to locate objects.

The constellation itself is compact, mainly consisting of a triangle of stars to the east of Vega and another pair of similar brightness to the south, toward Albireo in Cygnus.

Object	Type	Mv	Magnification	Distance
Epsilon Lyr	Double star	4.0	×200	160 light years
M57	Planetary nebula	8.8	×100	2300 light years

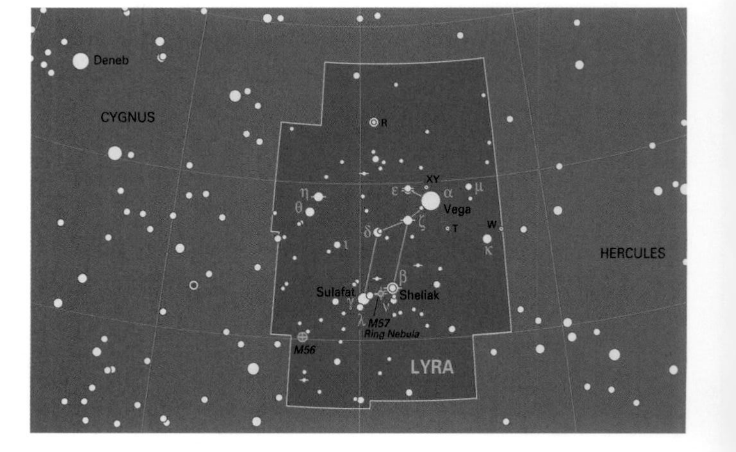

▲ *The components of each of the two doubles of the multiple star system Epsilon Lyrae are separated by 2.3 and 2.6 arc seconds, so they are a test for a 60 mm telescope.*

Beta Lyrae is an unusual variable star. It is an eclipsing binary (see page 75) with individual components so close together that each star is distorted into an egg shape. This is not visible in a telescope, but amateur astronomers can confirm it for themselves by monitoring the brightness changes of the star. It varies between magnitudes 3.3 and 4.4 in 12.9 days, and when its light curve is plotted, the results are much smoother than would be expected from a more normal eclipsing binary such as Algol.

Epsilon Lyrae, the Double Double

Young observers can easily see with the naked eye that the northernmost star of the triangle beside Vega is a double star. The two components are separated by 208 arc seconds, which is $3\frac{1}{2}$ arc minutes. Those with less keen eyesight can see this easily with binoculars or a finder scope. But look at Epsilon through any but the smallest telescope and you should see that the two components of the star are themselves double. The periods of the two double stars are 1000 and 600 years, so even over a lifetime of observing you will not notice any great change; nevertheless, the Double Double is a showpiece of the sky and features high on the list of objects to observe at star parties.

M57, the Ring Nebula

This planetary nebula is another favorite, not least because it is so easy to find, roughly midway between Beta and Gamma Lyrae. Point the telescope between these stars, possibly slightly nearer to Beta, and it is in the center of the field of view. However, you need a moderate power to see it well – around 100 is fine. It appears as a ghostly ring some 1.4 arc minutes across – about twice the size of Saturn as seen in the telescope. The central star is about 15th magnitude and you need a large amateur telescope to see it.

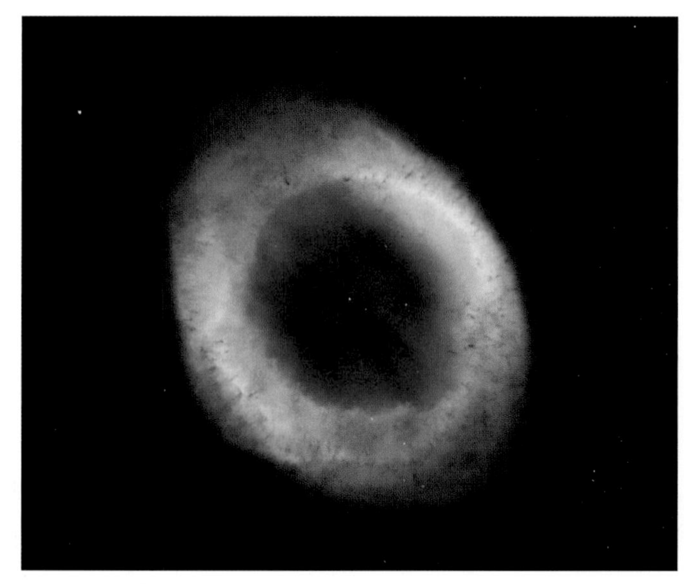

▲ *The best doughnut in the sky, the Ring Nebula displays strong colors in this Hubble Space Telescope photograph.*

Sagitta/Vulpecula

These two tiny constellations have a popularity beyond their size because of their location within crowded starfields of the Milky Way. The arrow shape of Sagitta, lying midway between Albireo and Altair, is easily found when the skies are dark enough to show the Milky Way properly, though binoculars are needed in bright skies. Vulpecula is less obvious but is easily located from Albireo.

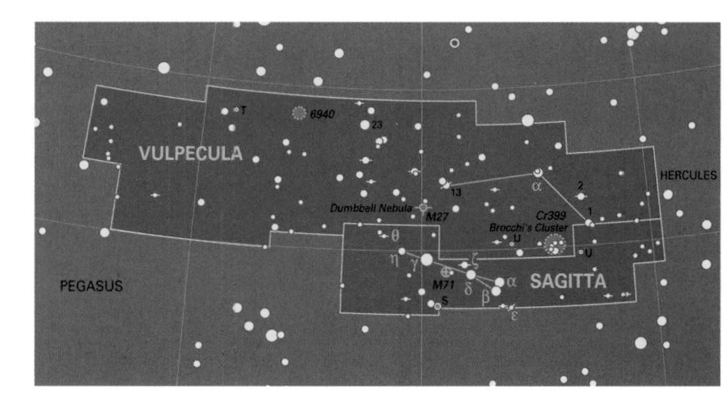

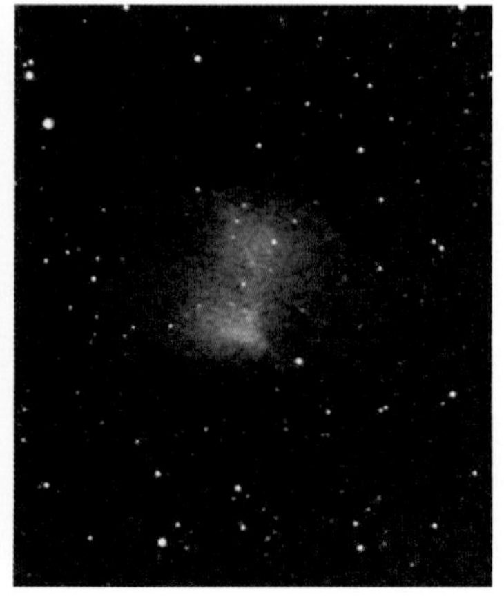

◄ *This view of M27 was taken with a digital camera on a 127 mm refractor, with an exposure of just 60 seconds.*

Planetary nebula M27, the Dumbbell Nebula

Even binoculars will show this bright and large planetary nebula. Though it is in Vulpecula, many observers locate it by using the stars of Sagitta – it is at a right angle to the arrow from the faint end star, Eta. The nebula itself is magnitude 7.3, though it appears fainter than a star of this magnitude because its light is spread over an area about a quarter of the size of the Full Moon. Its name comes from its double-lobed appearance in a telescope.

Object	Type	Mv	Magnification	Distance
M27	Planetary nebula	7.3	×50	1200 light years
M71	Globular cluster	8.2	×100	13,000 light years

▲ *Cluster M71 in Vulpecula is now generally considered to be a loose globular cluster rather than an open cluster.*

Open cluster M71

Point a telescope midway between Gamma and Delta Sagittae and you will find the cluster M71. This remote object seems to be midway between an open cluster and a globular cluster; astronomical authorities differ as to how it should be classified, but in a telescope it looks decidedly like a globular cluster as it is hard to resolve into stars.

The Coathanger

This well-known and attractive asterism gets its name from its shape, though it is also known as Brocchi's Cluster. The stars that comprise the asterism are all at different distances so it is nothing more than a chance alignment. The asterism is best viewed in binoculars.

▶ *There is no mistaking the Coathanger – its alignment of stars is unique.*

Aquila

Despite being a prominent constellation in the Milky Way, Aquila is surprisingly devoid of bright and well-known deep sky objects, though there are good starfields for binocular stargazing. Its brightest star, Altair, is easily identified as there are two less bright stars on either side of it. This trio lies at one side of a large diamond-shape of stars, with an offshoot to a C-shape of stars at the northern end of Scutum.

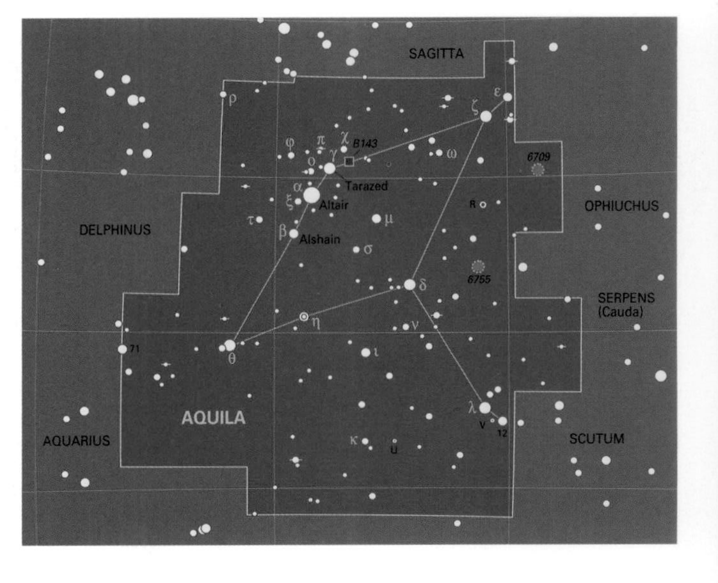

Variable star Eta Aquilae

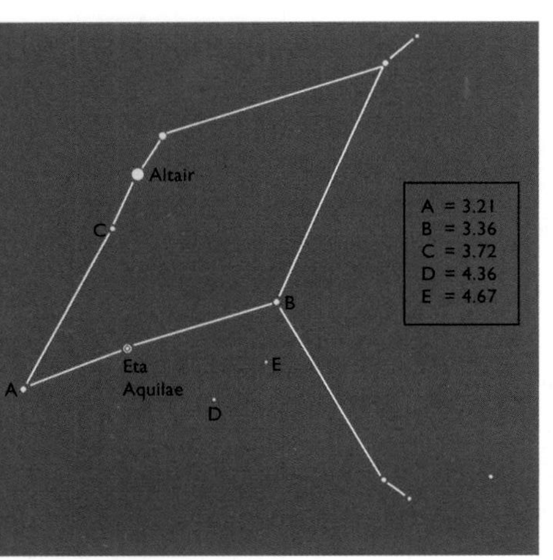

One of the brightest of the Cepheid type of variable star, Eta Aquilae is easy to find and to make estimates of because it is surrounded by convenient comparison stars. It varies between magnitudes 3.5 and 4.4 every 7.2 days. Members of the UK-based Society for Popular Astronomy have been observing this and other bright variable stars for many years.

◀ *Eta Aquilae, a variable star whose brightness is best estimated with the naked eye. Use these comparison stars to make your own estimates.*

Object	Type	Mv	Magnification	Distance
Eta Aql	Variable star	3.5–4.4	Naked eye	1200 light years

GROUP 4

Aquarius

At the dawn of history, when people in the Middle East were telling the stories about the stars that have come down to us in constellation names, they noticed that when the Sun was in the regions of the sky we now call Aquarius and Capricornus, the rainy season was upon them. The constellations in this area still have a watery connection, Aquarius being the Water Carrier who constantly pours water from his water jar into the ocean. The Water Jar itself is a group of four fairly faint stars in the shape of a sideways Y, but as with many asterisms it is the shape that is more distinctive than the brightness. To modern eyes the Water Jar and Alpha together resemble a jet plane more than a water jar. But there is a ragged line of more faint stars, which represents the water tumbling from the jar into the sea.

The brightest star in Aquarius is only slightly brighter than 3rd magnitude, and the whole area is rather barren of stars. One of the Messier objects shown on the map, M73, has frustrated astronomers for years as it is simply a group of four faint stars. Why it was included in the catalog remains a mystery. The constellation also contains two well-known planetaries and two globular clusters.

Object	Type	Mv	Magnification	Distance
NGC 7293	Planetary nebula	6.5	×10	450 light years
NGC 7009	Planetary nebula	8.0	×200	2400 light years
M2	Globular cluster	6.5	×50	50,000 light years
M72	Globular cluster	9.3	×50	55,400 light years

NGC 7293, Helix Nebula

You would expect the nearest planetary nebula to the Solar System, at magnitude 6.5, to be a fine sight, but in fact it is a great challenge to view, particularly for more northerly observers for whom it is rather low in the sky. The problem is that its brightness is spread over an area of sky more than a third of the diameter of the Full Moon. As a result it has a very low surface brightness and is easily lost in haze or light pollution. A clear, dark country sky and binoculars or a low-power telescope are needed. Even photography is a challenge.

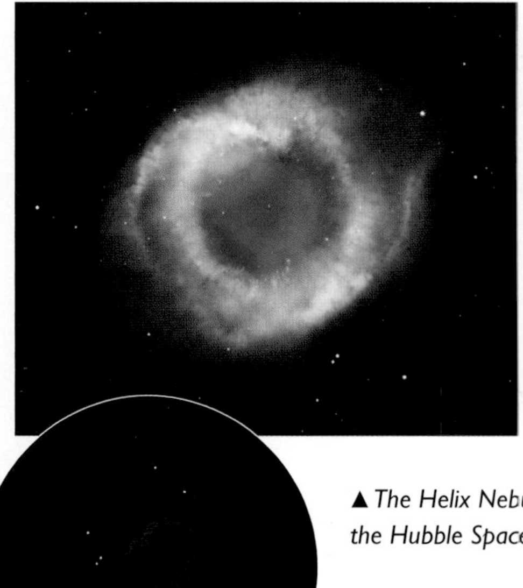

▲ *The Helix Nebula as seen by the Hubble Space Telescope.*

◀ *Viewed with a 200 mm reflector at ×72 from a suburban site near Brisbane, Australia, the Helix Nebula was only visible using an O III filter.*

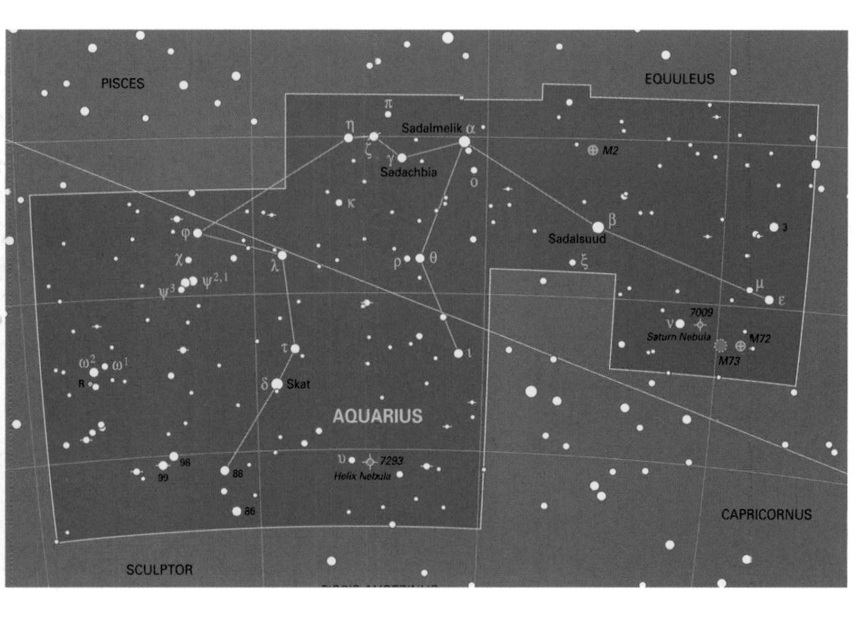

At first glance, a photograph of the Helix looks like the Ring Nebula in Lyra, but a closer look shows that it has two distinct rings of material arranged in a coil or helix seen end-on. This structure is hard to see visually, requiring large amateur telescopes as well as good skies and usually an O III filter.

NGC 7009, the Saturn Nebula

There is a great contrast between the Helix Nebula and the Saturn Nebula. The Saturn is also a planetary, and is less bright in total at magnitude 8. But it is small and its surface brightness is much higher than that of the Helix, so it is quite easily observable with small telescopes. The Saturn Nebula is about 1° due west of the 4th-magnitude star Nu Aquarii, making it simple to find. However, it is indistinguishable from a star unless you use a power of 50 or more.

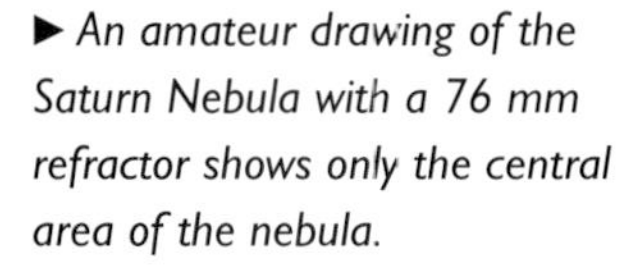

▶ *An amateur drawing of the Saturn Nebula with a 76 mm refractor shows only the central area of the nebula.*

With a small telescope the Saturn Nebula appears as a misty oval patch about the same apparent size as the planet Saturn. Only with a large telescope are projections visible on either side, making the resemblance with that planet much greater.

Globular clusters M2 and M72

It is fortunate that M2 is a bright object, at magnitude 6.5, otherwise it would be quite hard to locate in a small telescope, not being near any bright stars. Find it by locating Beta Aquarii and moving about 5° north, as the two objects have almost exactly the same right ascension.

In a small telescope M2 is a misty ball, and at around 50,000 light years it is more distant than many other bright globular clusters. This means that its individual stars are quite faint, so you will need a medium-sized telescope to see many stars in it.

Not far from the Saturn Nebula lies M72, which is even more remote. It is notoriously hard to resolve into stars.

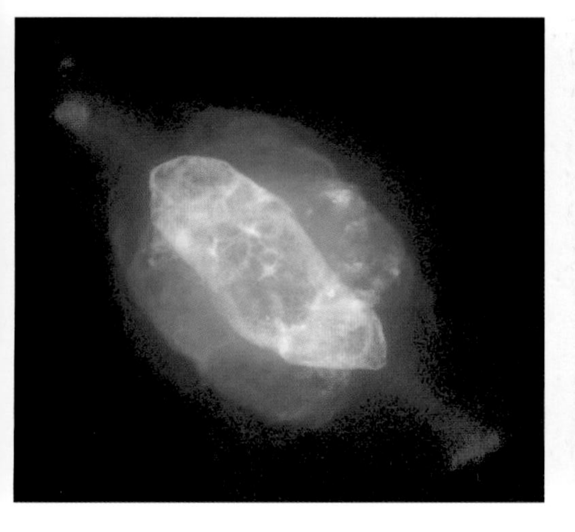

▲ *M2 as photographed using a 0.9 m telescope at Kitt Peak.*

◀ *The same instrument was used for this view of nearby M72.*

◀◀ *The extensions that give the Saturn Nebula its name are easily seen on the Hubble Space Telescope image at far left.*

Pegasus

This region of sky, well away from the Milky Way, is lacking in stars. So although the main stars of Pegasus are not particularly bright, they do stand out to form an easily recognizable square. You might think that the Square of Pegasus would be a compact box, but it is actually fairly large – getting on for about 20° across. Its simplicity makes it a classic signpost constellation for finding other stars – it will lead you to the stars of Andromeda, Pisces, Piscis Austrinus, Aquarius, and Cetus very easily. Pegasus is, therefore, an excellent starting point for learning the evening skies toward the end of the year, though it is rather lacking in bright deep sky objects. The Square itself is also rather barren; counting the number of stars within it is a useful means of comparing the eyesight of different observers, or of one observer on different occasions. Our chart shows 36 stars down to 6th magnitude, but keen-eyed observers in good skies may see more.

The mythological figure of Pegasus is a winged horse, though oddly the traditional depiction is upside down as seen from the northern hemisphere, and includes only the front half.

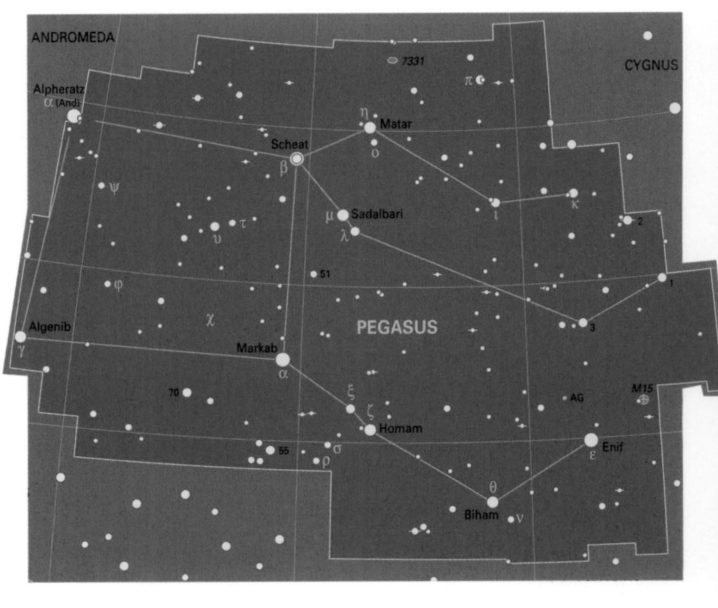

◄ *Globular cluster M15, photographed using a CCD on a 355 mm Schmidt-Cassegrain telescope.*

Globular cluster M15

The lone star Enif, well separated from the Square of Pegasus, is a good guide to finding M15 with binoculars or a finder. It is easily spotted with binoculars, and it makes an interesting comparison with the globular M2, which lies 13° to the south and slightly farther east. Both globulars are of similar total brightness, but M2 is more concentrated toward the center. However, unlike closer globulars, such as M13 or M22, M15 needs a medium-sized telescope to resolve it into stars.

Object	Type	Mv	Magnification	Distance
M15	Globular cluster	6.2	×50	33,600 light years

Capricornus

This is a large rough triangle of stars, most of which are lost in the murk from more northerly and light-polluted skies as it is fairly far south. The best way into Capricornus is to follow the line of three stars in Aquila containing Altair, going as far to the southeast as Albireo (Beta Cygni) is to the northwest. You come to the constellation's brightest star, Alpha Capricorni or Algedi.

The traditional constellation figure of Capricornus is a sea goat – though this is not a creature that features in any myths, unlike centaurs and lions. It has been linked with the Greek god Pan, who had goat's feet and who developed a fish's tail to escape a sea monster. Being in the ecliptic, it is often home to bright planets, which then outshine any star in the constellation.

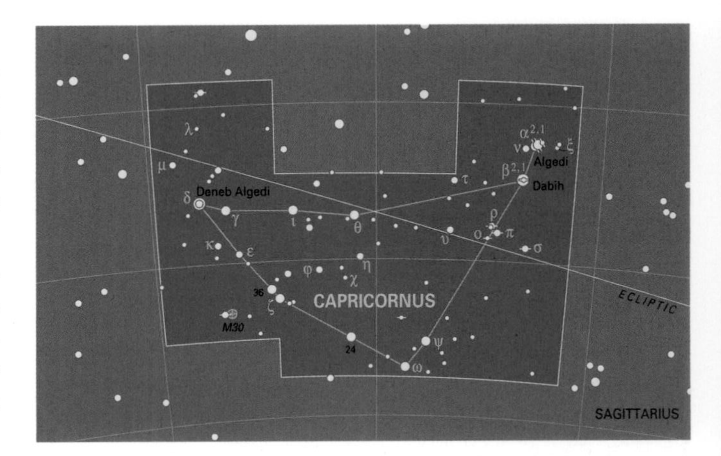

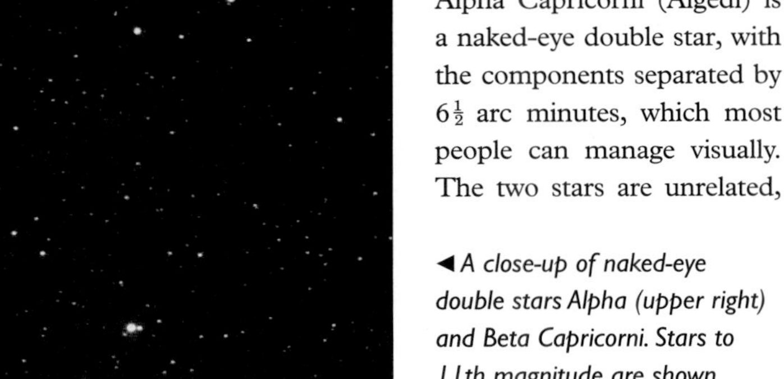

◄ *A close-up of naked-eye double stars Alpha (upper right) and Beta Capricorni. Stars to 11th magnitude are shown.*

Double stars in Capricornus

Alpha Capricorni (Algedi) is a naked-eye double star, with the components separated by $6\frac{1}{2}$ arc minutes, which most people can manage visually. The two stars are unrelated, however, with the fainter one being six times more distant than the brighter. The star is another candidate for the title "Double Double," like Epsilon Lyrae (see page 95), but unlike Epsilon Lyrae the secondary stars in each case are considerably fainter than the main stars.

Beta Capricorni is a double with stars of different colors – yellow and blue. The components are separated by $3\frac{1}{2}$ arc minutes, which puts the star well into the binocular category, though some may manage it with the naked eye.

Object	Type	Mv	Magnification	Distance
Alpha Cap	Double star	3.7	Naked eye	106/568 light years
Beta Cap	Double star	3.0	Naked eye	330 light years

Pisces

The sign of the fish continues the watery theme of this part of the sky. The constellation itself is particularly faint, however, and none of the stars is brighter than magnitude 3.7, which places virtually the whole constellation out of reach to the naked eye in light-polluted areas. In a good country sky, however, you can see that there are two lines of stars starting from Alpha Piscium (Alrescha) – one running between Andromeda and Aries, and the other below the Square of Pegasus, where it ends in a circlet of stars that marks the head of one of the fish. Being close to the celestial equator, Pisces is equally visible from both hemispheres.

Telescopically, this is another barren area, containing just one significant Messier object, the galaxy M74. Like many other face-on spirals this is notoriously hard to see, as it is 9th magnitude and quite large; some people can see it better in a good finder than in the main instrument.

Alpha Piscium (Alrescha) is a double star of roughly equal components separated by 1.9 arc seconds, making it a good test for a 60 mm refractor and more easily visible in a 100 mm or larger telescope.

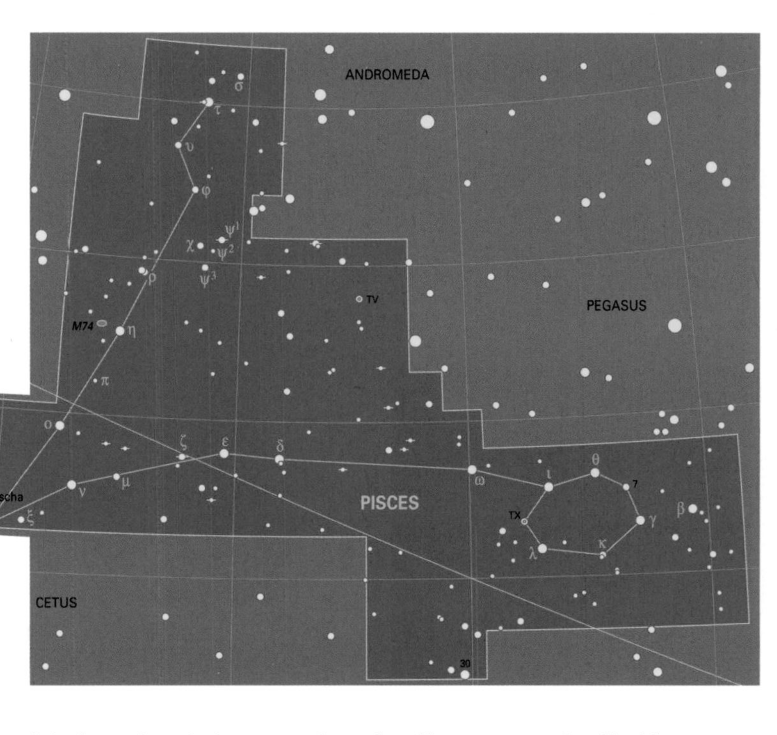

The Circlet asterism and TX Piscium

This evenly spaced circlet of stars just below the Square of Pegasus is easy to spot despite its faintness. Its brightest member, Gamma Piscium, is magnitude 3.9. But its greatest claim to fame is the irregular variable star 19 Piscium, also known as TX Piscium. This alias results from its being a variable star – variables in a constellation are given letters in a complicated sequence which need not concern us here. The star itself is famous for its strong color, which is easily visible in binoculars or a small telescope. It varies erratically between magnitudes 4.5 and 6.2, though sources differ as to its normal range and there is a lot of scatter in the estimates. Old books will tell you that it is a type N star, but these days it is known as type C, for carbon. TX Piscium is one of the brightest carbon stars, and it is the carbon in its atmosphere that filters out much of its blue light, resulting in a red star.

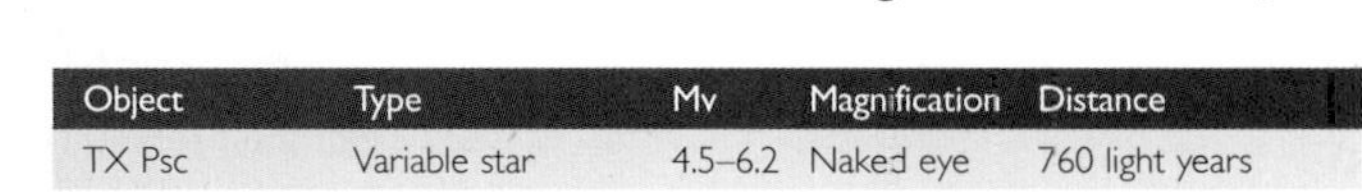

Object	Type	Mv	Magnification	Distance
TX Psc	Variable star	4.5–6.2	Naked eye	760 light years

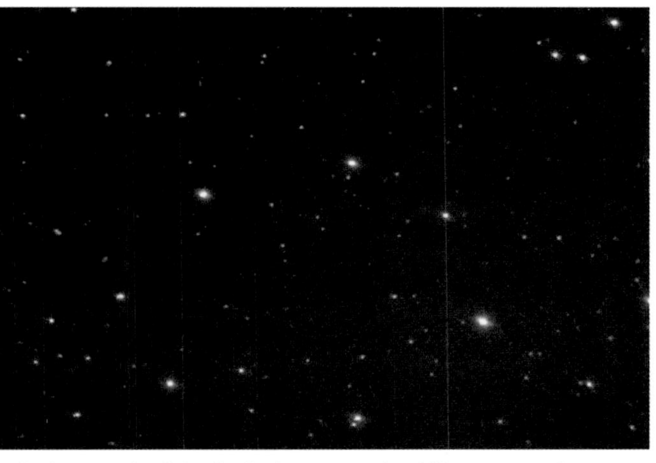

▲ *A photograph of the Circlet brings out the different star colors, notably that of TX Piscium at the left.*

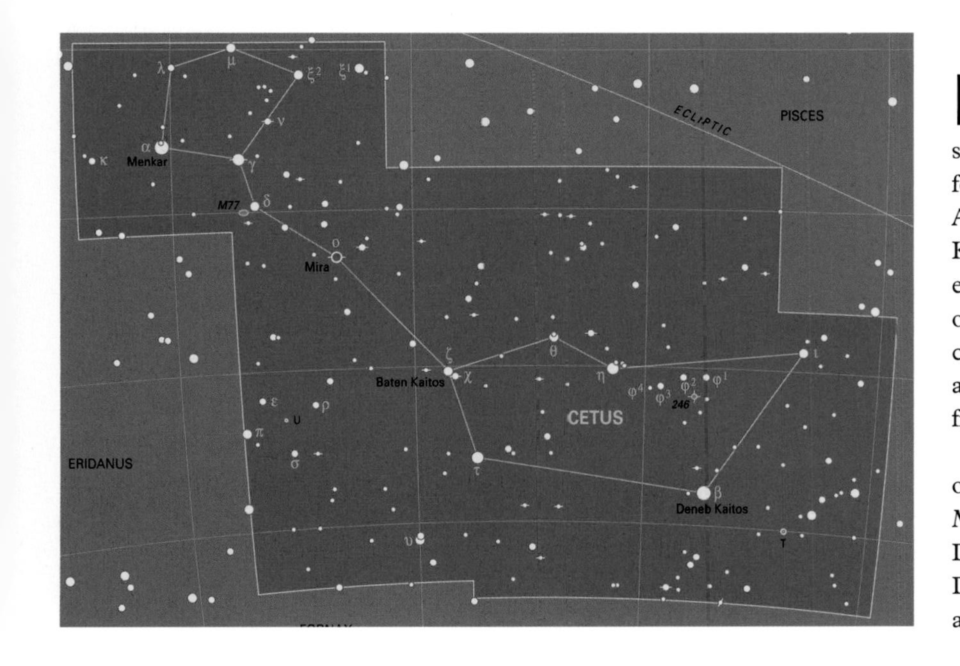

Cetus

Lurking below Pisces is the sea monster or whale, Cetus. Its head is a group of stars below Aries, and Alpha Ceti (Menkar) forms an equilateral triangle with Hamal in Aries and the Pleiades. Beta Ceti, Deneb Kaitos, is a long way south of the eastern edge of the Square of Pegasus, and is the only bright star in this part of the sky. The constellation itself straggles between Alpha and Beta, and is well placed for viewing from both hemispheres.

Cetus is not rich in bright deep sky objects – the most notable is the galaxy M77, which is conveniently located near to Delta Ceti and forms a neat triangle with Delta and 5th-magnitude 84 Ceti. M77 is a Seyfert galaxy, which is a type of spiral

◀ *The disk of the giant star Mira in visible light, as seen with the Hubble Space Telescope. The disk appears distorted, possibly as a result of starspots on its surface.*

galaxy with a bright nucleus. The active nuclei of Seyfert galaxies are thought to be low-power versions of quasars – highly luminous galactic nuclei that are powered by massive black holes. All quasars, however, are very remote, so Seyfert galaxies such as M77 allow us to glimpse some of the most energetic processes taking place in the Universe. The galaxy itself is 8th magnitude, and its condensed nucleus makes it easier to observe with small telescopes than many other Messier galaxies.

Object	Type	Mv	Magnification	Distance
Mira	Variable star	3–9	Naked eye	300 light years

Variable star Omicron Ceti (Mira)

As you follow your way down the line of stars from Alpha Ceti toward the tail of the monster, you may or may not come across one particular star – Mira. At times it can be easily seen with the naked eye, but at others it requires binoculars and a detailed star chart because it has faded to about 9th magnitude. Such behavior brought it to the attention of astronomers as early as the 17th century, at a time when changes among the stars were largely unknown. The name Mira means "the Wonderful," in honor of what were then thought to be amazing properties.

Mira has given its name to a whole class of stars that behave in the same sort of way, with considerable variations in brightness over a matter of months. The rise is fairly steep, so the star can appear almost nova-like by suddenly appearing in the sky over just a few weeks, reaching a maximum of about 3rd magnitude. Mira itself has a period of variability of about 330 days, so its maximum is around a month earlier each year. Its closeness to the ecliptic means that a maximum occurring between April and June will be unobservable because it will be lost in the twilight at the crucial time; for several years at a time Mira may always be below naked-eye visibility in the night sky.

Mira is one of only two stars whose actual disk has been imaged by the Hubble Space Telescope. Although it is about 300 light years away, it swells to such a size when at maximum that it is just within the HST's resolving power.

GROUP 5

Taurus

Although Taurus is a well-known and picturesque constellation, most of its stars are below 3rd magnitude. Yet it is one of the best-known patterns in the northern sky, and it bears some resemblance to a bull on a join-the-dots basis. Red giant star Aldebaran represents the eye of the bull, set within a V-shape of stars, the Hyades, which mark its face. Two outlying stars indicate the tips of its horns, while on its back sits the Pleiades star cluster. There is a suggestion that dots on a famous prehistoric cave painting at Lascaux, France, represent the Pleiades in exactly the same relationship to the head of a long-horned bull, and Taurus may be an even more ancient constellation than we realize. It is intriguing that the ancient cave painting shows just the front half of the bull, in just the same way as the more recent constellation depictions.

The most northerly extent of the ecliptic lies on the border between Taurus and neighboring Gemini, so the Moon and planets may be seen in Taurus and indeed can pass close to or even in front of Aldebaran and the stars of the Hyades and Pleiades. The constellation can be seen from both hemispheres.

Hyades star cluster

The stars of the Hyades are more widely spread than those of any other cluster, as it is the closest cluster to the Solar System, at a distance of just 150 light years. But Aldebaran, which seems embedded within it, is less than half that distance away. The individual stars are easy to count with the naked eye, with several at between magnitudes 3 and 5; in binoculars many more are visible.

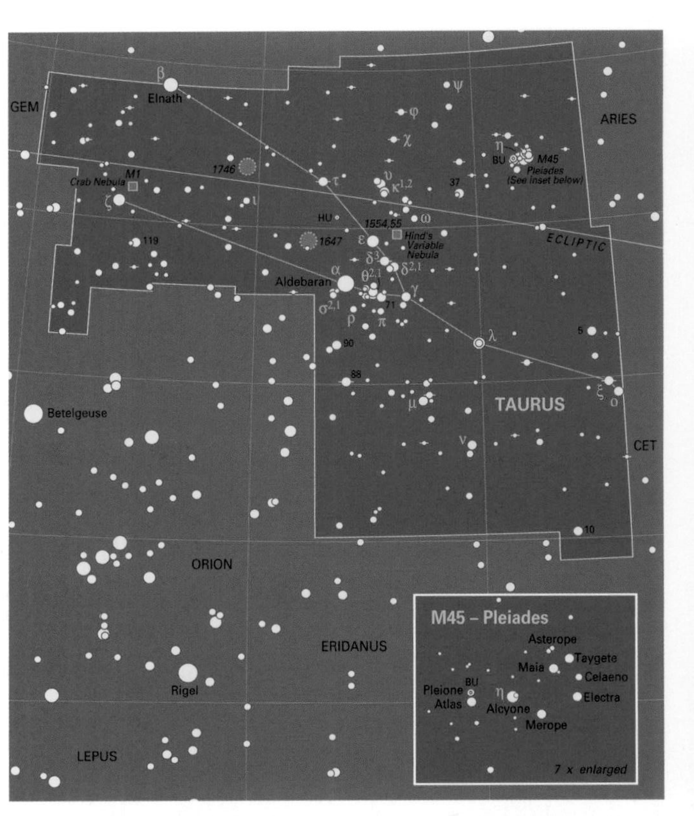

Object	Type	Mv	Magnification	Distance
Aldebaran	Star	0.9	Naked eye	65 light years
Hyades	Open cluster	–	Naked eye	150 light years
M45	Open cluster	1	Naked eye	370 light years
M1	Supernova remnant	8.4	×50	6300 light years

▲ *Aldebaran and the Hyades, with stars shown down to about 10th magnitude. The field of view here is 15°.*

M45, Pleiades star cluster

The Pleiades attract the eye like nothing else in the sky. The individual members of this brilliant cluster range from Alcyone at magnitude 2.9 down to very faint stars, and there are probably 500 or more in all. They have a total brightness of about 1st magnitude, and by averted vision the cluster is just as easily visible as a 1st-magnitude star, though appearing as a haze rather than a point of light. Much of the cluster is enveloped in a blue reflection nebula, known as the Merope Nebula after the star within its brightest part, though this requires binoculars or a low-power telescope in a good sky to be seen properly.

A popular name for the cluster is the Seven Sisters, and opinions vary as to how many stars are visible with the naked eye. Most people can certainly see six, and seven or eight are not hard, though some people can manage 14 or more. But the number seven was probably chosen for symbolic rather than numerical reasons. Although nine of the stars have names, the Pleiades were the seven mythological daughters of Atlas and Pleione, and their parents are represented by the two easternmost bright stars.

The cluster is a beautiful sight in any telescope, though a low power is needed for the best views. In a large telescope equipped with an ultra-wide-angle eyepiece in a dark sky the view is stunning, but even a city dweller will get a pretty sight with binoculars.

◀ *A long-exposure photograph of the Pleiades shows the Merope Nebula. Its blue color indicates that it is interstellar dust rather than gas. A short-exposure image of the Pleiades appears on page 76.*

M1, Crab Nebula

Long-exposure photographs of the Crab Nebula have made it one of the most famous astronomical objects, and every owner of a telescope will want to see it. But the reality is less dramatic than the photographs of an exploding star: a hazy oval is all that most people see. At 8th magnitude, the Crab Nebula is within reach of binoculars on a clear, dark night, but it has a low surface brightness and is easily wiped out by any light pollution or haze.

Fortunately, it is located close to a bright star, Zeta Tauri. The Crab Nebula lies just over a degree to the northwest of this star, about halfway toward a 6th-magnitude star that should be visible in a finder. M1 is about 8 arc minutes in width, which makes it larger than most planetary nebulae and galaxies, so a fairly low power is all that is needed. Few people see the claw-like filaments that gave the Crab its name, however, unless they are using large telescopes in very transparent skies.

The nebula itself is unique. It is a supernova remnant like few others in the sky, consisting of a shell of material kept glowing by radiation from a tiny central star. The supernova that created it was seen to explode in the year 1054 by Chinese astronomers and presumably by most others of the world's population, but only the Chinese definitely recorded it. It shone as bright as Venus, and was one of only four supernovae to have been observed in our Galaxy during the last millennium.

▲ *The Crab Nebula photographed using a 3.5 m telescope. It shows the dramatic filaments from the star that was seen to explode there over 950 years ago.*

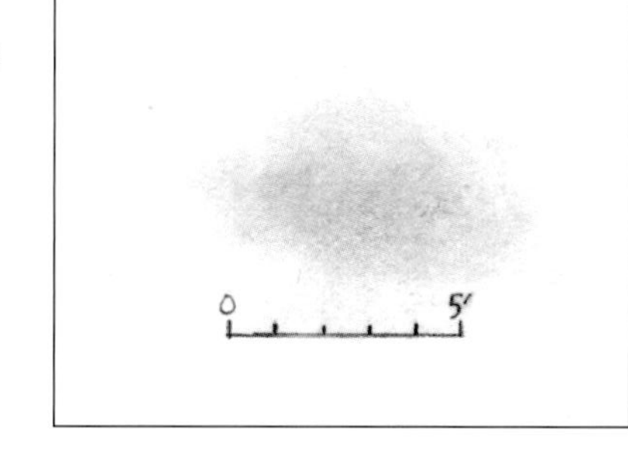

▲ *A drawing of the Crab made with a 76 mm refractor. It shows just the oval outline of the nebula.*

Auriga

Though not as well known to the general public as nearby constellations such as Orion or Taurus, Auriga is a favorite with astronomers because of its richness in interesting objects. It lies in the northern Milky Way so it is ideally placed for northern-hemisphere observers, but it is also accessible quite low down for most southern observers.

Auriga himself is a charioteer, though curiously he is saddled with a goat and her kids, which he carries in his left arm. The brilliant yellowish star Capella, sixth brightest in the sky, marks the goat and two of the adjacent triangle of stars mark the kids – known as the Haedi. The traditional pattern of stars is a pentagon, but the most southerly of these stars is actually spoken for by neighboring Taurus, where it has a respectable existence as Beta Tauri. So although the pentagon remains, there is no Gamma Aurigae and the star in question is exactly on the border between the two constellations.

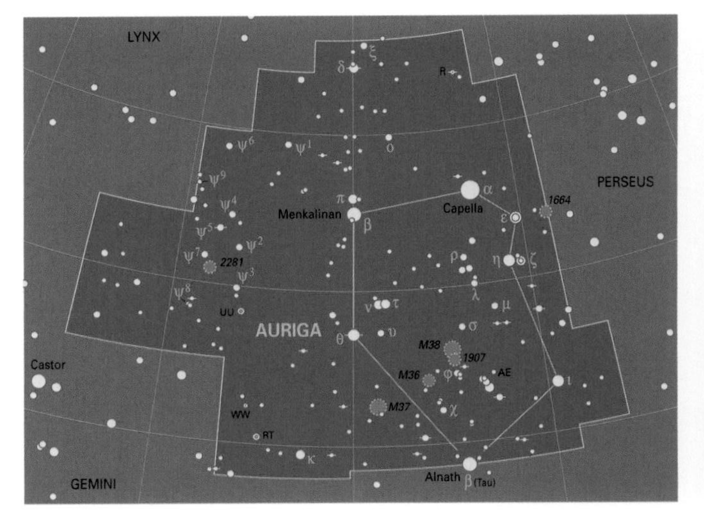

The Haedi and Epsilon Aurigae

Some books say that all three stars adjacent to Capella are the Haedi, but old constellation drawings show only two kids, which would be Eta and Zeta. It is a coincidence that two of the three stars are eclipsing binaries (see page 75) – a fairly rare type of variable star. Zeta is a common-or-garden eclipsing binary, with a period of 2.7 years. Its brightness drops from 2.7 to 3.0 for 40 days when its dimmer companion star partially hides it.

But Epsilon, the nearest of the three stars to Capella, is a different matter altogether. Its companion star takes 27 years to orbit it, and the eclipse reduces the star's magnitude from 2.9 to 3.8, which is quite noticeable, for over a year. Such a long eclipse can only be caused by an enormous object, far larger than any known star. The eclipsing body would have the same diameter as Saturn's orbit around the Sun, yet be invisible. Probably it is a cloud of dust, but there is great uncertainty. The last eclipse of Epsilon Aurigae lasted from 2009 to 2011.

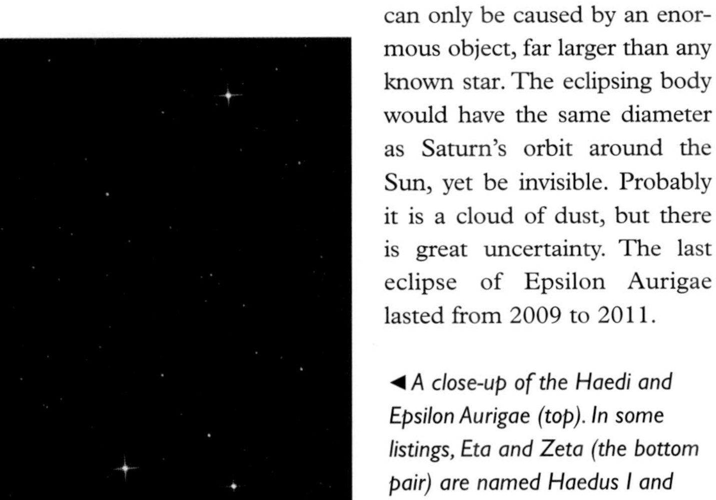

◀ *A close-up of the Haedi and Epsilon Aurigae (top). In some listings, Eta and Zeta (the bottom pair) are named Haedus I and Haedus II.*

Open clusters M36, M37, and M38

A line of open clusters adorns the center of the pentagon of Auriga. In one of the numerous quirks of astronomical nomenclature, they are numbered from east to west in the order M37, M36, M38. All three are at roughly the same distance of between 4100 and 4400 light years, but there are noticeable differences between them. All are easily visible to the naked eye in good skies and with binoculars in poorer skies, though a small telescope is needed to see the individual stars.

The richest of the three is M37, and it has inspired earlier authors to wondrous prose: "A magnificent object, the whole field being strewed as it were with sparkling gold-dust," according to the 19th-century observer Admiral Smyth. In a small- or medium-sized telescope it is certainly a beautiful sight, and it even shows through light pollution. M36 is a slightly poorer relation in terms of numbers of stars: it is smaller but with brighter and indeed younger stars. Nearby M38 is the largest of the three clusters, but less rich in bright stars. A smaller and fainter cluster, NGC 1907, lies nearby; it is just visible with small telescopes, consisting of stars mostly fainter than 11th magnitude.

Object	Type	Mv	Magnification	Distance
M36	Open cluster	6.3	×50	4100 light years
M37	Open cluster	6.2	×50	4400 light years
M38	Open cluster	7.4	×50	4200 light years
Flaming Star	Variable star	–	Photographic	1455 light years

AE Aurigae and the Flaming Star Nebula

Photographers of the M38 region may find that a nearby star has a red haze around it. This is the 6th-magnitude blue variable star AE Aurigae, which is illuminating a patch of hydrogen gas, known as the Flaming Star Nebula. AE Aurigae is an escapee from the region around the Orion Nebula, farther south, and is thought to have been flung away when a companion star exploded as a supernova, releasing AE Aurigae from its orbit. Visually the nebula is very difficult to see, though a hydrogen-beta filter may help.

◀ *The central part of Auriga, showing the three clusters M37, M36, and M38 running diagonally across a field of view of $6\frac{1}{2}^{\circ}$.*

▶ *The Flaming Star Nebula is to the right of a distinctive pattern of stars. The left-hand nebula is IC 410.*

Perseus

Few constellations have as much to offer as Perseus. This section of the northern Milky Way contains a range of interesting objects to suit all tastes.

In Greek mythology, Perseus was the hero of many adventures. But there is one particular story told in the stars – his rescue of the beautiful princess Andromeda from the sea monster Cetus, far to the south. In a previous episode he had slain the Gorgon Medusa, whose mere glance would turn a man into stone. He still carries her head hanging from his belt.

Surrounding Alpha Persei (Mirphak) is a glittering loose cluster of stars, known simply as the Alpha Persei Cluster. It is best viewed in binoculars, where it appears as a prominent S-shape of stars resembling a rollercoaster. Among other open clusters in Perseus is M34, which is easily resolved into stars with binoculars.

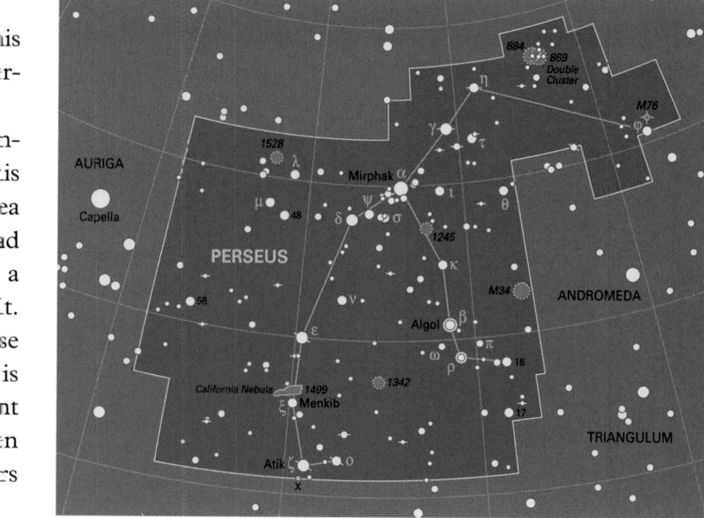

◀ *The Alpha Persei Cluster, also known as Melotte 20. Mirphak itself is at one end of a long S-shape of stars.*

Variable star Beta Persei (Algol)

The Gorgon's head is marked by a particularly famous star, Algol, whose name means "the Ghoul." Every 2 days 21 hours, this star winks – dimming from magnitude 2.1 to 3.4, and taking about 10 hours to do so. Algol was the first eclipsing binary to be identified, and stars of this sort continue to be called Algol-type stars.

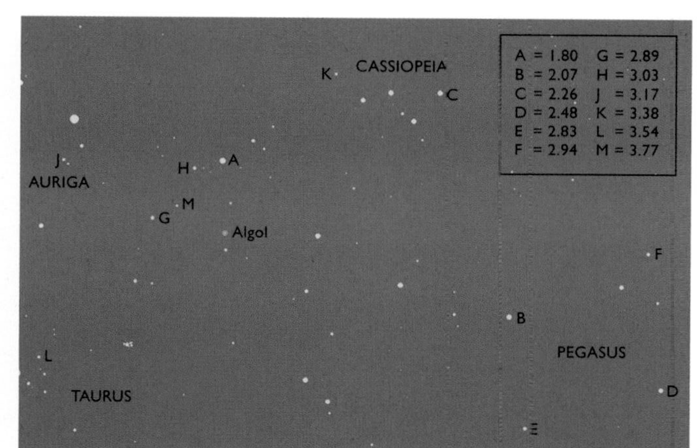

▲ *A comparison star chart, which can be used to follow Algol's variations. They are predicted in sky handbooks and are available online.*

Double Cluster, NGCs 869 and 884

To many observers this pair of clusters is as great a jewel of the sky as the Pleiades. Though they are by no means bright they are visible with the naked eye, by averted vision at least, in a good sky midway between Perseus and Cassiopeia. In binoculars they are a delight, and in a telescope under good conditions they can be breathtaking. One of the clusters by itself would give most others a run for its money; but side by side they are unbeatable. Yet these are not nearby clusters – they are both over 7000 light years away, getting on for twice the distance of M37, say. It is the sheer number of faint stars that makes them impressive, and if they were much closer they would be an amazing sight. There are several red stars in NGC 884, which add to the spectacle.

▲ *The Double Cluster, NGC 884 (left), also known as Chi Persei, and NGC 869 (right), also known as h Persei.*

Planetary nebula M76, the Little Dumbbell

In theory, the Little Dumbbell should be a great challenge to owners of small telescopes because it is claimed to be the faintest object in the Messier Catalogue. In practice, however, this 10th-magnitude planetary has a fairly high surface brightness so it is a good deal easier than some objects that are supposedly much brighter. M76 is simple to locate once you have found 4th-magnitude Phi Persei, which is easier to find by star-hopping from Gamma Andromedae than

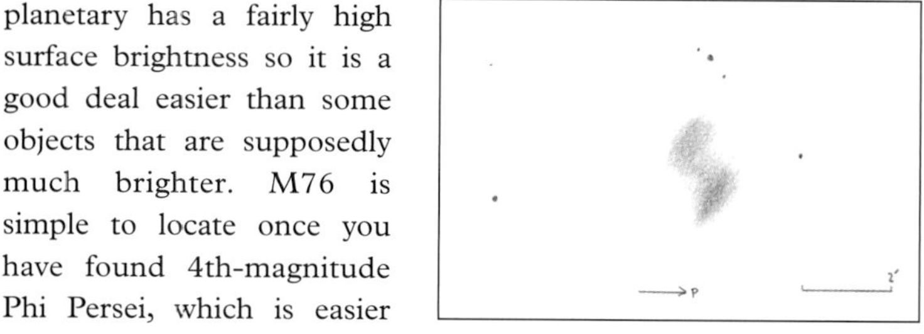

▲ *A drawing of M76 made using a 460 mm Dobsonian telescope.*

Object	Type	Mv	Magnification	Distance
Alpha Per	Star	1.8	Naked eye	500 light years
Beta Per	Variable star	2.1–3.4	Naked eye	90 light years
NGC 869/884	Open cluster	4.3/4.4	×10	7100/7400 light years
M76	Planetary nebula	10.1	×100	3400 light years
NGC 1499	Nebula	–	Photographic	1000 light years

from the stars of Perseus. M76 is just to the west of a 6th-magnitude star a degree north of Phi.

► This close-up of the California Nebula has east at the top. Xi Persei is the blue star to the right of the nebula.

NGC 1499, the California Nebula

One for the photographers, this patch of pink nebulosity near Xi Persei looks at first sight like a rather fat pink banana. But turn it with east at the top, and it becomes a fair representation of the state of California. It is surprisingly easy to capture using fast film and shows through moderate light pollution, though DSLRs will struggle; visually, however, it requires dark skies and an O III or hydrogen-beta filter. The nebula is a hydrogen cloud that just happens to be illuminated by Xi Persei.

Gemini

The two leading stars of Gemini – Castor and Pollux – are not a matched pair, but they have been regarded as twins since ancient times. Mythologically they are the sons of Leda, Queen of Sparta, but to two different fathers – a process that does not normally result in twins. The two stars do have subtly different colors, Castor being bluish and Pollux yellowish. A good way to remember which is which, once you know the sky a little, is from their capital letters: Castor is closest to Capella while Pollux is closest to Procyon.

Two loose lines of stars mark the figures of the twins themselves. Their feet dabble in the Milky Way, where there are several interesting deep sky objects. The ecliptic also runs through Gemini, so its pattern may be altered by the presence of a planet. Although Gemini is a northern constellation it can be seen from both hemispheres – though for southern-hemisphere observers it is quite low.

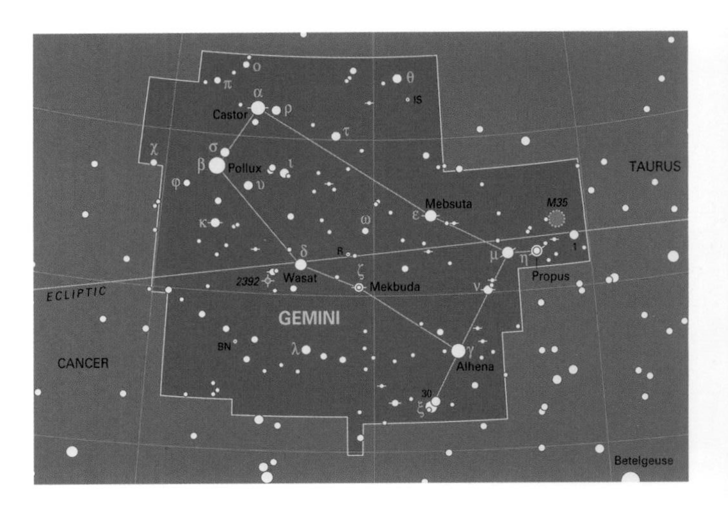

Multiple star Alpha Geminorum (Castor)

Despite being labeled Alpha, Castor is the fainter of the twins. Nevertheless, it consists of no fewer than six individual stars, of which three are visible using small telescopes. The two main components are over 4 arc seconds apart, and a 60 mm telescope should easily separate them. Look carefully and you should see a 9th-magnitude star about 60 arc seconds from the main pair. Each of these three stars is a close double star, though the components cannot be separated visually, so Castor is six stars in one.

▼ Mu and Eta Geminorum and open cluster M35, showing stars down to 11th magnitude.

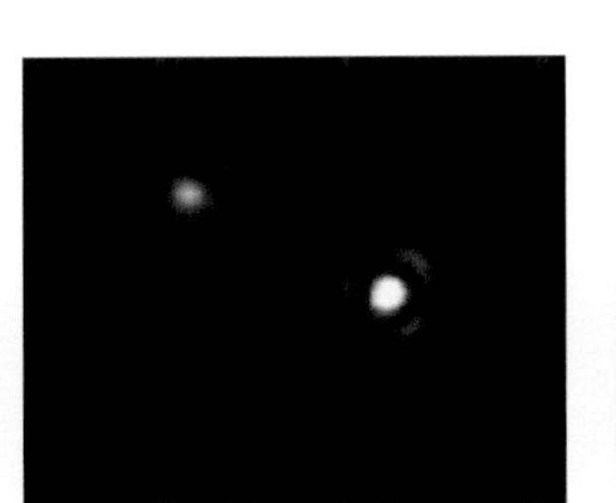

▲ Castor as photographed through a 280 mm telescope is split into two stars of magnitudes 1.9 and 2.9.

Open cluster M35

The two stars Mu and Eta Geminorum are easily spotted at the western end of Gemini. They act as signposts to the cluster M35, just to their northwest. M35 is visible to the naked eye in good conditions or binoculars in poorer skies, and is one of the larger open clusters in the sky, so binoculars will easily show some individual stars. A telescope reveals masses of stars from magnitude 9 downward.

NGC 2392, the Eskimo or Clown Face Nebula

One of the brighter planetary nebulae, the Eskimo is fairly easy to locate from a little triangle of stars near Delta Geminorum, Wasat. With a small telescope you can see it looking like an 8th-magnitude star, which with some magnification becomes an oval shell with maybe a blue-green tinge. With a large telescope and high magnification, some detail becomes visible, with the outer shell looking a little like a fringe surrounding a face, hence the nebula's name.

► A drawing of the Eskimo Nebula made with a 460 mm Dobsonian telescope from a suburban location.

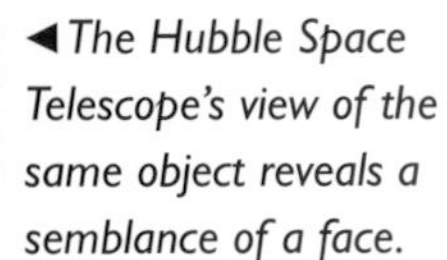

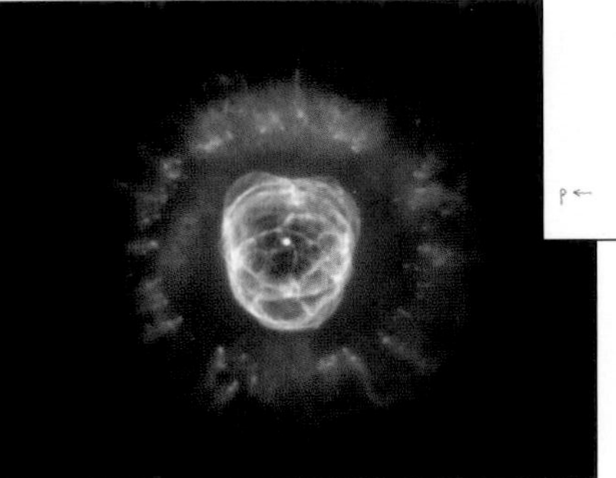

◄ The Hubble Space Telescope's view of the same object reveals a semblance of a face.

Object	Type	Mv	Magnification	Distance
Castor	Double star	1.6	×200	52 light years
M35	Open cluster	5.3	×25	2800 light years
NGC 2392	Planetary nebula	9.2	×100	3000 light years

GROUP 6

Orion

There is no constellation more brilliant than Orion. When it is in the sky, it burns through any light pollution and haze, and being slap-bang on the celestial equator it is visible the world over. In the northern hemisphere its stars are symbolic of the winter sky; in the southern hemisphere they mean hot summer nights.

The three stars of Orion's Belt, equally spaced in an almost straight line, have no counterpart anywhere in the sky, and provide an unmistakable signpost. The area is also the closest stellar nursery to Earth, and within its borders are good examples of stars at virtually every stage in their lives. Most of the bright stars we see in Orion share a common origin in the starbirth region, though they have subsequently spread over a wider area of space.

The figure we call Orion is a hunter, who lives in Greek legends. Before that he was a giant in the folk tales of other cultures. He faces a raging bull, Taurus, and holds up a shield, marked by the line of stars Pi^1 to Pi^6 Orionis, made from the skin of a lion that he has slain. From his belt hangs a glowing sword which includes the Orion Nebula. At his heels are two dogs, the constellations Canis Major and Canis Minor, while below him in the sky cowers a hare, the constellation Lepus.

In a good dark sky the whole area seems to glow, and long-exposure photographs show that it is indeed a mass of faint nebulosity. To its east is a great glowing arc, known as Barnard's Loop. This is not readily visible to the eye, but is easily photographed on fast film. It is probably the remnant of a supernova, a star that has exploded at the end of its life. This may seem odd if Orion is a stellar nursery, but the most massive stars, the sort that end up as supernovae, go through their life cycles in a matter of maybe 10 million years – so they are exploding while other stars in the region are still being born.

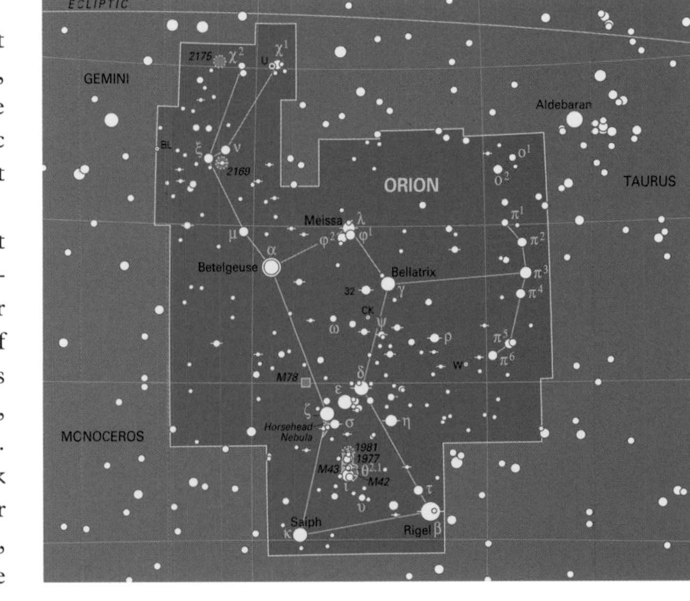

▲ *This photograph of Orion was taken with a diffusing filter over the lens to enhance the star colors. The difference between Betelgeuse and the other main stars is clear.*

Alpha Orionis (Betelgeuse)

Although Betelgeuse (pronounced "Bet-el-jooz" rather than "Beetlejuice") is the Alpha star of Orion, it is usually noticeably fainter than Beta Orionis, Rigel. However, it is a variable red supergiant, and can vary between magnitudes 0.2 and 1.3 or maybe even fainter. This is a wider brightness range than any other bright star, and it is worth making an estimate using the chart on page 75 whenever you observe.

Betelgeuse's red color is very obvious in comparison with the other main stars in Orion, which are mostly bluish. It is considerably closer, at 500 light years compared with the estimated 1500 light years of the main Orion starbirth region, and there are suggestions that it may not be part of the Orion complex. It will someday become a brilliant supernova in its own right, but this is not likely for millions of years.

Sigma Orionis

This is a delightful multiple star. Even a small telescope will immediately show three stars, two being of 6th magnitude alongside the main 3rd-

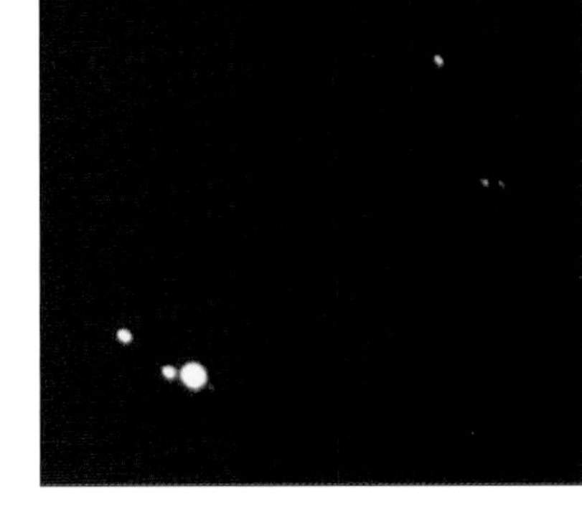
► *Sigma Orionis is one of the most celebrated double stars in the sky, with a number of blue stars in the same field.*

Object	Type	Mv	Magnification	Distance
M42/M43	Nebula	4.0/9.0	×7	1600 light years
Horsehead	Dark nebula	–	×100	1600 light years
Flame	Nebula	–	Photographic	1000 light years
Trapezium	Multiple star	–	×100	1600 light years

magnitude star. If you look more closely and with a higher magnification you should see that on the opposite side of the main star is a closer 9th-magnitude companion. The main star itself is also a close double, though this is a difficult object. Other nearby stars add to the spectacle.

M42 and M43, Orion Nebula

▲ *The Orion Nebula, M42, with M43 just above the main nebula and NGC 1977 at the top.*

What we see as the Orion Nebula is really only the tip of the iceberg. The whole of the Orion region is occupied by a great dark cloud of hydrogen and molecules that hides more distant regions of the Galaxy. Only the region where stars are actually forming and illuminating the cloud is visible. Radio and infrared measurements show that there are many other stars within the cloud that are hidden from view.

The Orion Nebula rivals the Eta Carinae Nebula as the brightest gaseous nebula in the sky, and it is bright enough to be visible with the naked eye as a misty star in a good sky, and with binoculars in a light-polluted area. This is a favorite target for all astronomers, as it offers not only a visual spectacle and a suitable subject for sketches, but also an unbeatable area for photography. Although photographs on film show it as red, digital cameras and the eye are not so red-sensitive and they tend to show it as whitish. Some people see the central regions as greenish, because there is also a strong green component to the light.

The region is wreathed with nebulosity, and the outer wings of the nebula contain delicate details visible even in very small telescopes. A dark central bay in the nebula is popularly called the Fish's Mouth from its shape.

At the end of the Fish's Mouth lies the multiple star Theta[1] Orionis. This consists of stars that have been born from the nebula within the past million years or so. Although you are unlikely to spot a new star popping into view, changes do take take place in this area from time to time. The four main stars of Theta are known as the Trapezium from their shape, but there are at least two other 10th-magnitude stars in the pattern visible with medium-sized telescopes using high magnification. The Trapezium is simply the brightest part of a cluster of maybe 1000 stars, many of which are visible only in infrared, which is not absorbed as much by the gas and dust in the area as is the visible light.

To the north of the Fish's Mouth lies M43, which would be a popular target for observation in its own right if it were not overshadowed by the larger and brighter M42. Farther north still, surrounding a group of stars of which 42 Orionis is the brightest, is the fainter reflection nebula NGC 1977. Go yet farther north and you find a small cluster of stars known as NGC 1981.

Horsehead Nebula and Flame Nebula

The Horsehead Nebula is undoubtedly one of the top ten illustrations in astronomical books, but as a visual object it is elusive. Those who live under pitch black skies might disagree, but for most people it is lost in the murk. This does not prevent them from looking, and there is great satisfaction in actually locating this famous object. It usually requires a large telescope, though it has been glimpsed in medium-sized instruments.

The Horsehead is actually a dark nebula within a strip of nebulosity called IC 434 running south of Zeta Orionis. To the east of Zeta lies a brighter part of the nebula, known as NGC 2024, which is also called the Flame Nebula from its ragged appearance in photographs. A hydrogen-beta filter is a considerable help in seeing the faint glow of IC 434 and the bay within it that we call the Horsehead. In fact, this type of filter is often referred to as the "Horsehead Filter" because this is one of the few objects for which it is actually useful. Others include the California Nebula in Perseus and the Flaming Star Nebula in Auriga.

Although these objects are a challenge to visual observers, they are fairly easy targets for astrophotographers with fast, red-sensitive film and a telephoto lens on a driven equatorial mount. Exposure times of 10 minutes or so should show them quite easily.

► *Hubble's close-up of the Trapezium area shows many of the stars in the cluster. Interesting tails can be seen on stars that are in the process of being formed; they are caused by strong gas outflows.*

► *In this view with north at left, the bright star to left of center is Zeta Orionis. Below it is the Flame Nebula, and to its right is IC 434, with the Horsehead in the middle of the nebula. The star at top is Sigma Orionis.*

Canis Major

The celestial tableau centered on Orion also includes two dogs, larger and smaller. Canis Major is the larger dog, and its brightest star, Sirius, is the brightest in the night sky. It hardly needs instructions to be able to find it – the three stars of Orion's Belt point southeastward directly to it, and it is visible the world over. Canis Major contains a number of other bright stars that even without the help of Sirius would make it a major constellation. Its stars make the shape of a rather odd stick dog, which lies just to one edge of a rather faint part of the Milky Way, and is accompanied by Lepus, the Hare. The constellation of Lepus suffers from being so close to its illustrious neighbors as it contains some relatively bright stars in its own right.

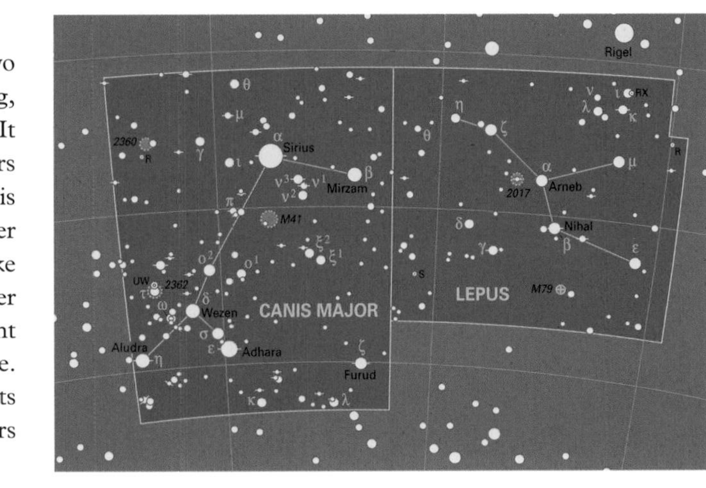

Alpha Canis Majoris (Sirius)

From our point of view, Sirius is top dog in our little corner of the Galaxy, outshining the Sun by more than 20 times. But as stars go, it is not particularly special: it is only bright because it lies so close to us – a mere 8.6 light years, just twice the distance of the nearest bright star, Alpha Centauri.

▲ *Sirius as viewed in close-up with the Hubble Space Telescope. Sirius B is the tiny dot to the upper left of the main star image. The spikes are caused by diffraction of light in the optical system and by CCD artefacts.*

Exactly why Sirius is regarded as the Dog Star is unknown. It may have been called this long before the constellation itself got its name. To the ancient Egyptians it was a crucial star, as its appearance in the dawn sky before sunrise heralded the vital flooding of the Nile. During northern-hemisphere summer it is in the sky at the same time as the Sun, which is why hot, sultry summer days are known as Dog Days, when the heat of Sirius is supposed to be added to that of the Sun.

Object	Type	Mv	Magnification	Distance
Sirius	Star	–1.44	Naked eye	8.58 light years
M41	Open cluster	4.6	×10	2300 light years

Sirius twinkles like no other star, merely because it is so bright. It can even give rise to UFO scares, with turbulence in our atmosphere making it appear to be sending out rays and sparks.

Accompanying Sirius in the sky is another star, a tiny white dwarf known as Sirius B. It is not a faint star, at 8th magnitude, but its closeness to Sirius makes it very hard to spot without using special techniques. This is the closest of this unusual class of stars, which have diameters comparable to those of planets but virtually as much mass as a star, making them exceedingly dense.

Open cluster M41

Just 4° south of Sirius is the bright open cluster M41, which is easily visible with binoculars and can be resolved into stars with a low magnification on a telescope. It is made all the prettier by three or four orange stars of 6th or 7th magnitude.

▲ *A view of M41 taken with a 0.6 m telescope at Kitt Peak.*

Canis Minor and Monoceros

The constellation of Canis Minor, the lesser dog, is noted for containing the star Procyon, but little else. Like the greater Dog Star, Sirius, Procyon is close, at 11.4 light years, and with our Sun, Alpha Centauri, and Sirius it makes a little group within the Galaxy that we might consider our own neighborhood. Curiously, Procyon has a white dwarf companion like Sirius, though fainter and even more difficult to see.

Sirius, Procyon, and Betelgeuse in Orion are sometimes called the Winter Triangle, a description that applies only in the northern hemisphere, to match the Summer Triangle of Vega, Deneb, and Altair.

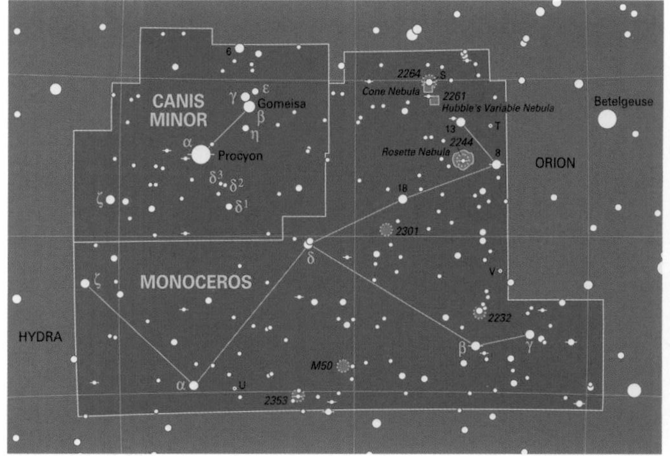

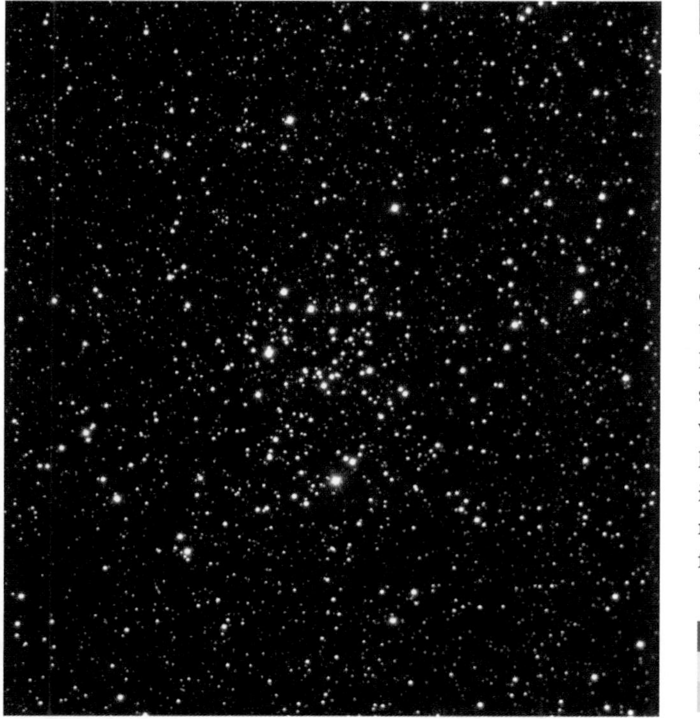

▲ *Open cluster M50 taken with a 0.6 m telescope at Kitt Peak.*

Monoceros is not a well-known constellation, and many stargazers would be hard pressed to identify its pattern of 4th-magnitude stars. This Unicorn inhabits the region to the east of Orion and south of Canis Minor. Though it may be unimpressive as a constellation, it lies within the Milky Way and therefore contains a number of interesting objects. Beta Monocerotis, which with Gamma lies between Sirius and the Belt stars of Orion, is a beautiful triple star with three bluish stars.

Open cluster M50

As Monoceros is poor in stars, this cluster is probably easiest to find by locating Gamma Canis Majoris, just to the east of Sirius, and then moving 7° north. The cluster is easily visible in binoculars as a blur, and small telescopes will show a group of stars of which the brightest are 8th and 9th magnitude, with outliers that make it hard to decide on the boundaries of the cluster.

Rosette Nebula

Another of the sky's top book illustrations, the Rosette is very easy to photograph. Though its location is readily found, all that most people can see visually is a small cluster of stars known as NGC 2244, more scattered and less visually impressive than M50. But a telephoto lens and fast, red-sensitive film will yield a result with an exposure of minutes, though a longer exposure and good guiding are needed for really impressive results. The circular appearance of the Rosette makes it look like a supernova remnant, but it is an H II region from which the cluster of stars has been born.

Object	Type	Mv	Magnification	Distance
M50	Open cluster	6.5	×50	3000 light years
Rosette	Nebula	–	Photographic	5500 light years

▶ *The Rosette Nebula is a spectacular sight in photographs and records easily on red-sensitive film.*

GROUP 7

Vela

In Greek times this area was occupied by an enormous ship, Argo Navis – the very same ship in which Jason and the Argonauts set sail. But it proved to be too cumbersome for celestial navigation, so it has now been divided up into its component parts – the Sails (Vela), the Keel (Carina), and the Stern or Poop (Puppis). This was done after the stars had been assigned their Greek letters, so each constellation now has letters missing. Vela, for example, has no Alpha or Beta. Its brightest star, Gamma Velorum, however, has at least two claims to fame. It is a wide double star visible as a pair in a telescope at low power, and the brighter component is a particularly hot type of star known as a Wolf-Rayet star. Such stars have a temperature of about 60,000°C, compared with the Sun's modest 5700°C, and Gamma Vel is the brightest example.

Vela is regarded as a southern-hemisphere constellation, though it is visible from the southern United States. Like many southern constellations, it lacks a real pattern, and some of the stars making up the polygon that traditionally marks the sail's outline are unlettered and no brighter than the stars which are not part of the outline. Vela is on the fringe of the Milky Way in an area that is rich in stars of 4th and 5th magnitude.

Open cluster NGC 2547

Just 2° south of Gamma Velorum lies this open cluster, which can be located with binoculars. It is quite large so requires a low power in a telescope, and consists of a wide scattering of stars mostly of 8th magnitude and fainter.

◀ *NGC 2547 is a widely scattered star cluster best seen with binoculars.*

Object	Type	Mv	Magnification	Distance
NGC 2547	Open cluster	4.7	×25	1960 light years
NGC 3132	Planetary nebula	9.4	×100	2000 light years

NGC 3132, the Eight-Burst Nebula

A fine planetary nebula even for small telescopes, NGC 3132 is bright enough to be visible from light-polluted areas. The name comes from the loops within it, which are visible on photographs. Visually it appears as an oval disk somewhat larger than Jupiter, with a prominent central star. This star, however, is not thought to be the one responsible for the nebula. That honor goes to another much fainter object, visible on the spectacular Hubble Space Telescope photograph, very close to the apparent central star. NGC 3132 is also sometimes called "the Southern Ring Nebula" because of its similar easy visibility, though the ring is not as pronounced as with M57, the Ring Nebula in Lyra.

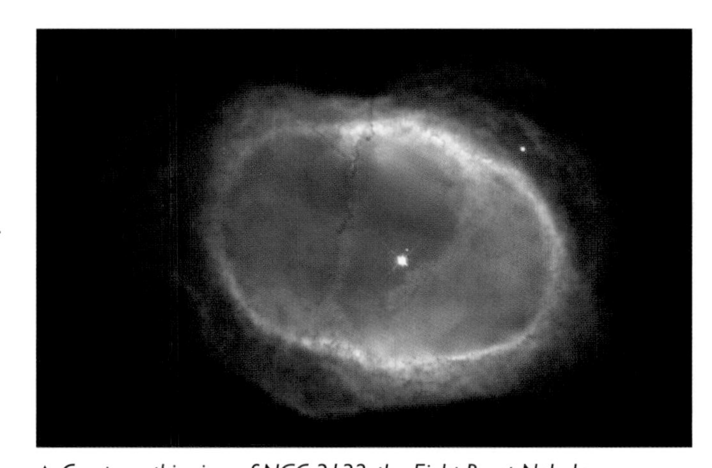

▲ *Compare this view of NGC 3132, the Eight-Burst Nebula, photographed with the Hubble Space Telescope, with the amateur photograph that is shown on page 78.*

Carina/Volans

The Keel of the ship Argo occupies one of the richest areas of the southern Milky Way. Like Vela, it is crowded with stars, and Carina includes the second-brightest star in the sky, Canopus, which is at the extreme western end of the constellation. You really need to be south of the Equator to see this part of the sky in its full glory. Volans, the Flying Fish, is a comparatively recent addition to the sky.

False Cross asterism

Vela and Carina have together created much confusion by forming, between them, a rival to the Southern Cross. This False Cross is actually larger and a somewhat different shape, and in particular it fails to indicate south as does the true Cross. Once you have seen the Southern Cross it is unmistakable, so it

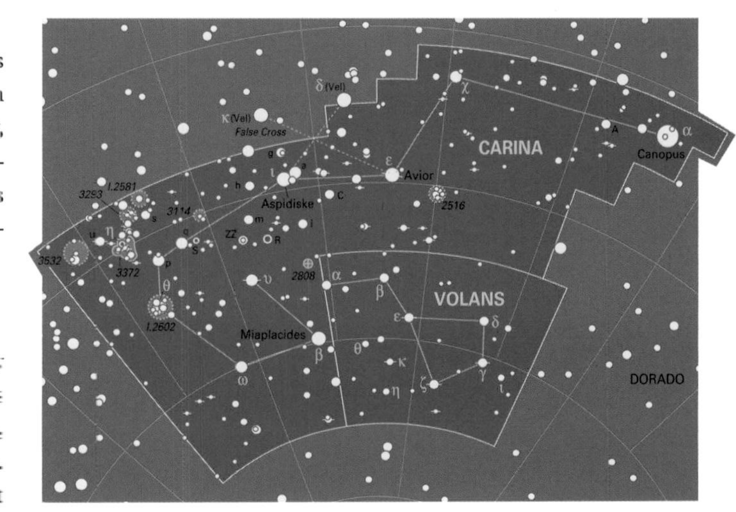

▲ The "False Cross" of stars in Vela and Carina is different in many details from the real one (see page 123).

is unlikely that many have been seriously misled by these competitors – but it is worth recognizing the difference to prevent yourself being caught out. The stars in question are Kappa and Delta Velorum, and Iota and Epsilon Carina.

Eta Carinae Nebula

The nebula that surrounds the star Eta Carinae is the largest and brightest in the sky. It lies in a star-studded region, and the whole area attracts the eye straight away, even in poor skies. It lies on the opposite side of the Southern Cross from Beta Centauri, at virtually the same declination, seemingly balancing the star in the sky.

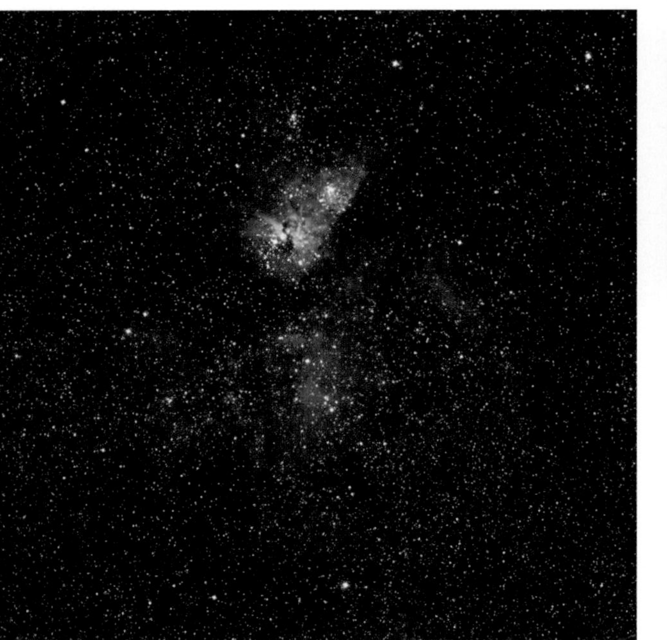

▲ The Eta Carinae Nebula. The star Eta itself is in the middle of the wedge-shaped nebulosity, to the left of the Keyhole Nebula.

Binoculars reveal a large hazy area with numerous stars, while in a telescope at low power there is a wealth of detail in the nebula, notably an L-shaped dark lane. The nebula is a starbirth region and numerous young stars and little clusters are visible surrounding the nebula.

The star Eta Carinae itself is currently fairly insignificant, at 6th magnitude. It lies within a dark zone of the nebula known from its shape as the Keyhole, though this patch has varied somewhat since it received its name in the mid 19th century. Eta is noticeably orange and is embedded in a bright nebulosity that is visible at high magnification. In 1843 Eta astonished observers by flaring up to magnitude −1, outshining even Canopus. It may do so again. Recent observations with the Hubble Space Telescope indicate that it is actually a double, with one very massive component at 100 to 150 times the mass of the Sun, and the other at a mere 30 to 60 solar masses.

Object	Type	Mv	Magnification	Distance
Eta Carinae Nebula	Nebula	1.0	×10	10,000 light years
NGC 2516	Open cluster	3.8	×25	1300 light years
NGC 3114	Open cluster	4.2	×10	3000 light years
NGC 3532	Open cluster	3.0	×25	1300 light years
IC 2602	Open cluster	1.9	×25	479 light years

Open cluster NGC 2516

Within a binocular field of Epsilon Carinae lies this splendid open cluster, visible with the naked eye under good conditions. Its brightest star is a 5th-magnitude red giant. Binoculars show many of its brighter stars and a haze of the fainter ones that are visible with a telescope.

▲ Open cluster NGC 2516, as seen on the Digital Sky Survey made using large Schmidt telescopes. The field of view is 1°.

► A wide-field view of NGC 3114, covering 4° of sky and showing stars fainter than magnitude 14.

Open cluster NGC 3114

Sweep just to the west of the Eta Carinae Nebula with binoculars and this beautiful cluster appears. It is a naked-eye object in good skies, easily resolved into stars with binoculars or a small telescope.

Open cluster NGC 3532

Every star cluster is different, and this one, celebrated as the richest open cluster, is neither the brightest nor the largest. To the naked eye it is a misty patch northeast of Eta Carinae. It lies in the same binocular field of view as Eta, and appears as a mass of stars of 7th magnitude and fainter. You will need a low power and a wide-field eyepiece on a telescope to do it justice, as it covers nearly 1°. The cluster includes several bright orange stars of type K, and there is a curious darker band across its center where there are few stars.

◀ *Binoculars will give a view of open cluster NGC 3532 similar to this photo, which has a field of 3.5° and shows stars to magnitude 10.*

Open cluster IC 2602

Nicknamed "the Southern Pleiades," this cluster, like the Pleiades themselves, is a test of eyesight – there are seven or eight stars visible with the naked eye, spread over a similar area to the Pleiades. However, this cluster's stars are in general fainter. Like the Pleiades, it is best seen in binoculars because of its large size.

▲ *This view of "the Southern Pleiades" shows a similar area to that of the original Pleiades, as seen on page 101.*

Puppis

This most northerly part of the old Argo Navis extends close enough to the celestial equator for much of its area to be visible to northern observers. The brightest parts, however, are really best seen from the southern hemisphere. Puppis lacks bright stars and a clear pattern. Its most readily recognizable section is a group of stars, which includes its brightest star, Zeta Puppis or Naos, to the southeast of Canis Major. There are, however, numerous scattered stars of 3rd and 4th magnitude.

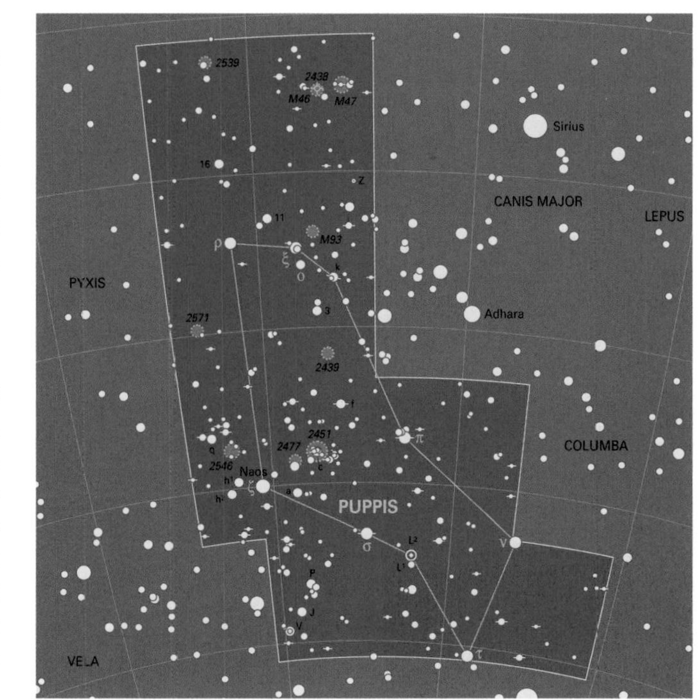

Open clusters M46 and M47

Follow a line from Beta Canis Majoris through Sirius and you arrive at this pair of Messier objects. The brighter of the two, M47, is visible with the naked eye under good conditions, though most people will need binoculars in average skies. The fainter M46 is also said to be a naked-eye object, but if there is light pollution around it can be a challenge even through binoculars. The pair make an interesting contrast – both cover a similar area of sky, but M47 contains less numerous but brighter stars, while M46 is more suited to a low-power telescope view as it has numerous stars of about 10th magnitude.

Open cluster NGC 2451

Another naked-eye cluster, NGC 2451 lies midway between Zeta and Pi Puppis, and surrounds a 3rd-magnitude orange star. Though good in binoculars, it lacks fainter stars so the telescopic view is less impressive.

▲ *The open clusters M46 (left) and M47 compared. The field of view is 5°, similar to that of binoculars.*

Object	Type	Mv	Magnification	Distance
M46	Open cluster	6.0	×10	5400 light years
M47	Open cluster	5.2	×10	1600 light years
NGC 2451	Open cluster	2.8	×7	850 light years

▲ *A Digital Sky Survey view of open cluster NGC 2451. Though there are faint stars across the field, most of the cluster members are of magnitudes 5 or 6.*

Scorpius

It is widely agreed that Scorpius is one of the few constellations in which you can join the dots and get a reasonable representation of the animal it is supposed to be. For many northern observers, however, the resemblance is lost because a crucial part of the constellation, the curve of the scorpion's sting, lies forever below the horizon. Antares marks the heart of the creature, with its claws extending westward, while its fearsome sting dips southward. Notice that the astronomical and astrological names for this constellation are subtly different – to astronomers it is Scorpius, not Scorpio. Capricornus is a similar trap for the unwary!

Scorpius is a very ancient constellation, and Greek mythology relates that it is the creature that killed the proud hunter Orion. To this day, Orion avoids the scorpion and his stars set as those of Scorpius rise.

The borders of the constellation include parts of both the Milky Way and the ecliptic, so occasionally the planet Mars can be seen very close to Scorpius' brightest star, Antares. The name Antares means "Rival of Mars," and they do have similar ruddy colors. Antares has a 5th-magnitude companion star just 2.6 arc seconds away – so close that at least a 75 mm telescope is required to see it. It is a type B star, but is often described as green. In theory green stars should not occur, so this is probably a contrast effect.

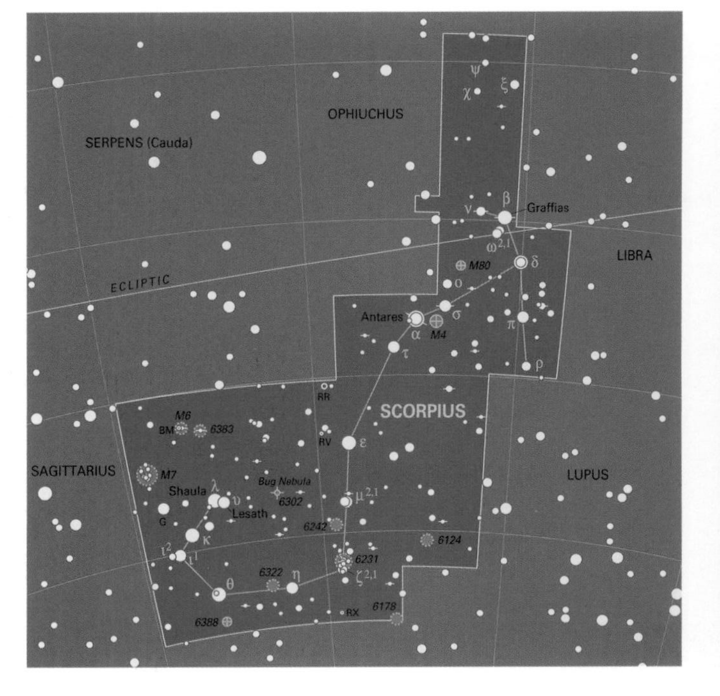

Globular cluster M4

This is a bright object very close to Antares, but much larger and fainter than other globular clusters. Find it by centering Antares in the field of view, then moving just over 1° westward. M4 may not be easy to find in a hazy sky because it is not a particularly rich or condensed cluster. You will need a small telescope to begin to resolve it into stars.

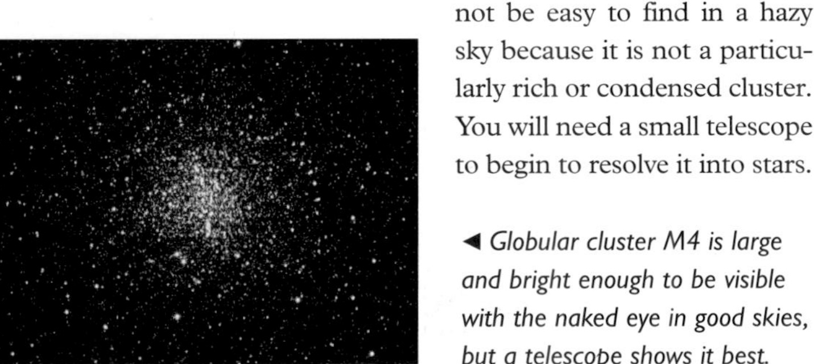

◀ *Globular cluster M4 is large and bright enough to be visible with the naked eye in good skies, but a telescope shows it best.*

Open clusters M6 and M7

▲ *M7 is in a richer part of the Milky Way than M6. It was mentioned by the ancient Greek astronomer Ptolemy, so it is sometimes referred to as Ptolemy's Cluster.*

Of several bright open clusters in Scorpius, these are the most prominent. Both are visible with the naked eye as misty patches and can be resolved into stars with binoculars. M7 has the brighter stars, while M6 is more condensed but richer in faint stars. In a low-power telescope of any size both clusters are impressive. Some people refer to M6 as the Butterfly Cluster because the pattern of the stars resembles a butterfly's open wings.

Open cluster NGC 6231

This is a large and bright cluster at the bottom of the sting of the scorpion, at the point where it turns eastward. It is somewhat smaller than the Pleiades, and its stars are less bright, so it is not particularly obvious to the naked eye. But with binoculars it is a brilliant sight. Even its pattern of stars is reminiscent of the Pleiades, but on a smaller scale.

◀ *This Kitt Peak photo of open cluster M6 suggests the butterfly-like arrangement of stars.*

▶ *Scorpius carries a jewel in its sting – the open cluster NGC 6231. Many of its stars are type O, and are only about 3 million years old.*

Object	Type	Mv	Magnification	Distance
M4	Globular cluster	5.6	×25	7200 light years
M6	Open cluster	4.2	×10	1600 light years
M7	Open cluster	3.3	×7	800 light years
NGC 6231	Open cluster	2.6	×25	5900 light years

Lupus and Norma

The stars of Lupus lie between Antares and Alpha Centauri. None of them is brighter than 2nd magnitude but the region is in a rich part of the Milky Way and offers good sweeping with binoculars. These constellations are mostly below the horizon for northern observers, except those in the southern United States or south of the Mediterranean.

Norma is undistinguished in terms of bright stars, but it contains some nice open clusters, as well as a bright patch of the Milky Way known as the Norma Star Cloud. A long rift in the Milky Way that runs through Lupus and Norma is referred to by Native Australians as "the Emu." The bird's head is the Coalsack in Crux, while the clouds in Lupus and Norma are its body. The emu's legs are hidden underneath its body as it crouches on its nest.

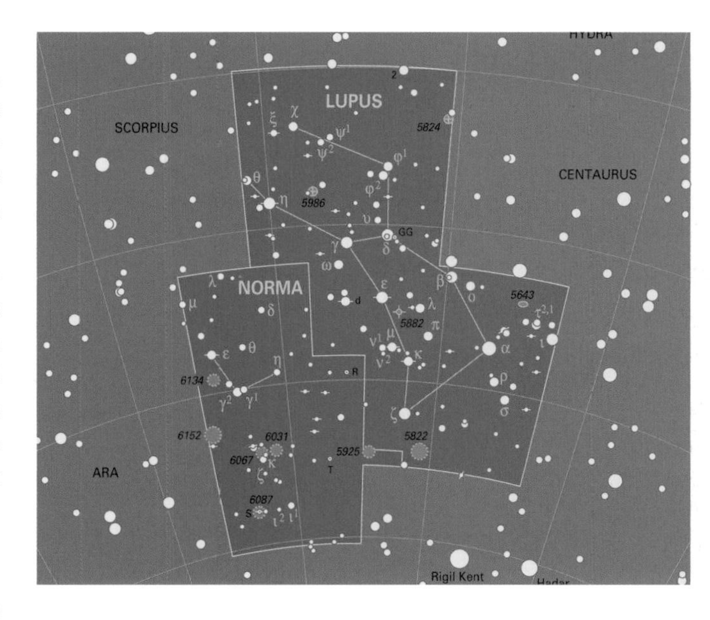

Open cluster NGC 5822

Some clusters, like this one, spring into view with binoculars when the sky is dark enough, but are otherwise difficult to see. It appears as a large diffuse haze with binoculars, while in a telescope you need a wide-field eyepiece because the cluster covers about 40 arc minutes and fills the view with stars. There are around 150 stars, all of around 10th to 11th magnitude. You will find the cluster about three-quarters of the way between Alpha Centauri and Zeta Lupi.

Open cluster NGC 6067

Having found NGC 5822, move exactly 10° eastward and you come to this rather brighter cluster. It is easily found in binoculars as a fairly condensed misty patch, and many 8th-magnitude and fainter stars are visible with a telescope.

Open cluster NGC 6087

Within the same binocular field as NGC 6067 lies NGC 6087, 4° farther south and 1° east. It is a more condensed cluster, with several 8th-magnitude stars and many fainter ones, some being yellow or orange when seen through a telescope. There is a double star near the center.

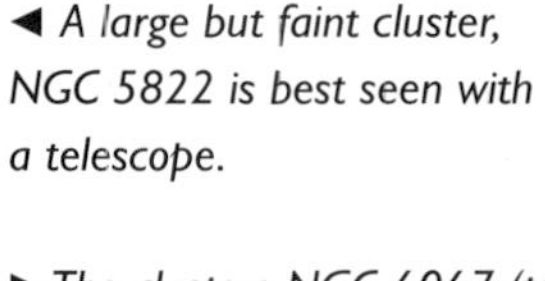
◄ *A large but faint cluster, NGC 5822 is best seen with a telescope.*

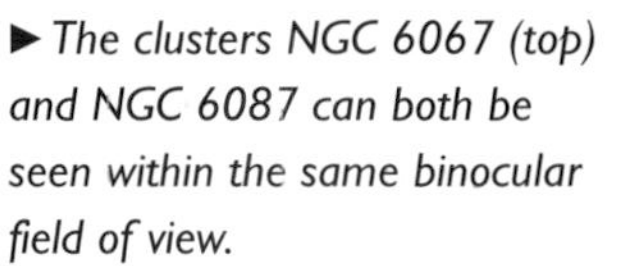
► *The clusters NGC 6067 (top) and NGC 6087 can both be seen within the same binocular field of view.*

Object	Type	Mv	Magnification	Distance
NGC 5822	Open cluster	7	×25	3000 light years
NGC 6067	Open cluster	5.6	×50	4600 light years
NGC 6087	Open cluster	5.4	×50	2900 light years

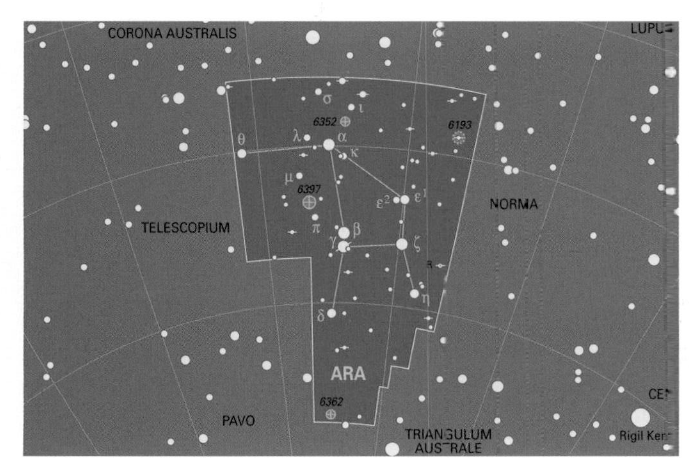

Ara

To the south of the sting of Scorpius lies this small and easily identified constellation, consisting of two slightly curved lines of stars. It is on the fringes of the Milky Way and contains several open clusters that are visible with binoculars, NGC 6193 being the brightest. Despite being fairly far south, the constellation was named in ancient times and may represent the altar on which Greek gods swore an oath of allegiance.

Object	Type	Mv	Magnification	Distance
NGC 6188	Nebula	–	Photographic	3800 light years
NGC 6193	Open cluster	5.2	×50	3800 light years
NGC 6397	Globular cluster	5.9	×25	7200 light years

◄ An amateur-taken image of NGC 6397. Because of its closeness to Earth, this cluster has been studied using the Hubble Space Telescope.

► A small telescope shows the sparse cluster NGC 6193, while the nebulosity NGC 6188 is a good photographic target.

Open cluster NGC 6193 and nebula NGC 6188

Though it is easily found in binoculars, the cluster NGC 6193 is not particularly noteworthy, as it has few bright stars. A couple of other open clusters – NGC 6200 and NGC 6204 – are visible within the same binocular field to the north. But this area comes into its own in long-exposure photographs, being wreathed in nebulosity with dark lanes and bays somewhat reminiscent of the Horsehead Nebula.

Globular cluster NGC 6397

Situated at the apex of a triangle made by Alpha and Beta Arae, this is one of the closest globular clusters to the Solar System, if not the closest. Being quite large, it is an easy binocular object, but a telescope and a little bit more power are needed to resolve it into stars. It is fairly loose and open, like M4 in Scorpius.

GROUP 9

Virgo

The Virgo region of sky is unimpressive to the eye, or even to binoculars. But to astronomers it is one of the most important areas of the whole heavens, for it contains the great Virgo Cluster of galaxies. But the constellation itself occupies a much wider part of the sky, and it is the second-largest constellation, after its neighbor to the south, Hydra.

The brightest star is Spica, a Latin name meaning "Ear of Wheat," which is what the maiden carries in her hand. The Virgin referred to is not the Christian one, but instead either the Greek goddess of justice or the corn goddess. The ecliptic runs the length of Virgo, so a bright planet may often rival Spica for attention. Virgo straddles the celestial equator, so is equally visible from either hemisphere.

This part of the sky is well away from the Milky Way, so it is comparatively barren in stars. This is fortunate, as it means that the Virgo Cluster is unobscured by the dust and gas clouds that would otherwise restrict our view. Our own group of galaxies, which also includes the Andromeda Galaxy and the Magellanic Clouds, is regarded as a minor outlier of the Virgo Cluster, the center of which lies some 50 million or so light years away. It contains around a thousand galaxies in all.

► A photomosaic of the central region of the Virgo Cluster, with M87 at center left and M86 and M84 at upper right.

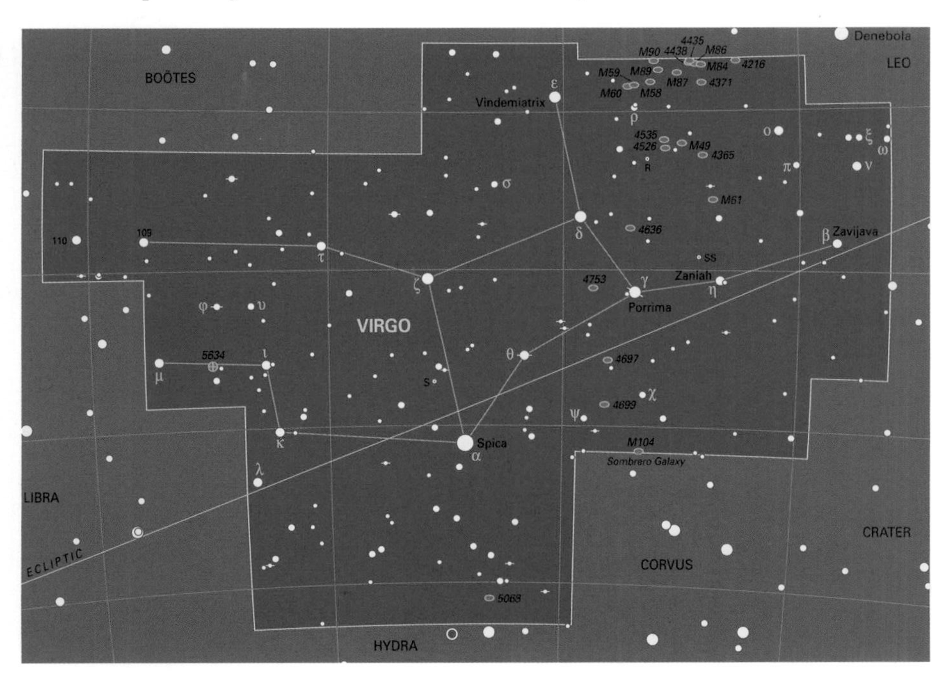

Exploring the Virgo Cluster

A star map makes the Virgo Cluster look as if one could hardly fail to stumble upon galaxies, but in fact they are not so easy to find, as the individual galaxies are quite small and faint. In a dark, clear country sky, the brighter members are just visible with binoculars if you know exactly where to look. Small telescopes will show the brighter galaxies if conditions are dark enough, as they are mostly 8th and 9th magnitude, but finding them can be a challenge. Owners of small Go To instruments may think that they will have no problems, but they may well find that unless their instrument is accurately aligned and the skies are truly dark, they will be rewarded with blank areas of sky, or uncertainty as to which object is which.

With medium and large telescopes on a good night, however, the situation changes.

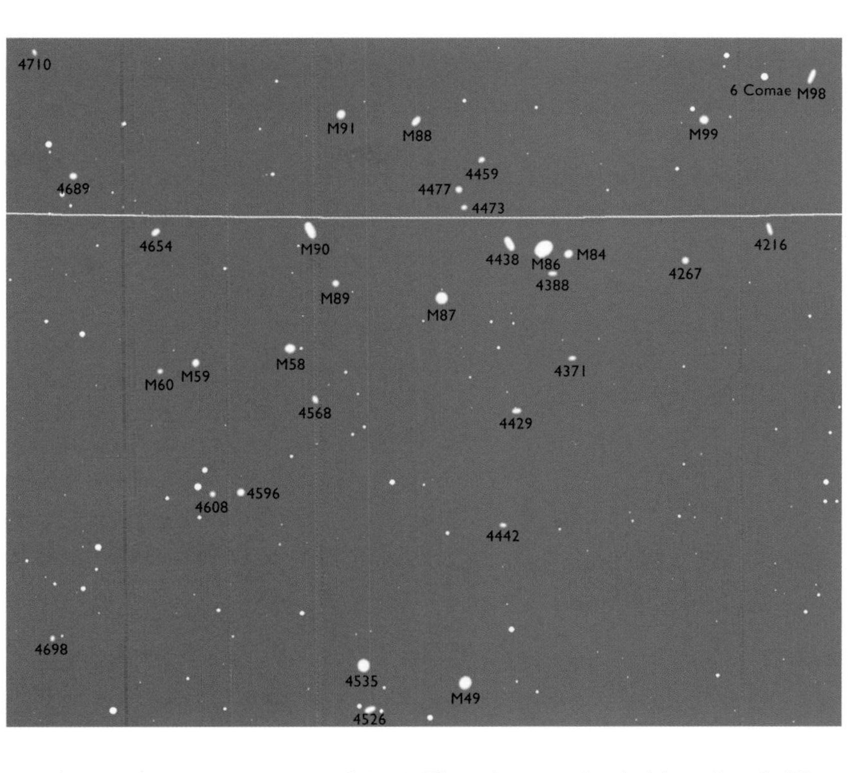

► *The Virgo Cluster mapped with stars to magnitude 9.5 and galaxies to magnitude 11, adapted from the SkyMap computer mapping program.*

The galaxies are all there, and you can hunt for fainter members to your heart's content. Simply pointing the instrument in the right direction will start to bring them in. But even so, if you want to be able to identify the galaxies and not get lost, you need to learn a route.

Of several such routes, possibly the simplest is to start with the star Denebola in Leo in the finder and then move eastward to find the 5th-magnitude star 6 Comae Berenices, $6\frac{1}{2}°$ away. As the map shows, you are now on the edge of the cluster (more properly known as the Virgo–Coma Cluster because it extends into Coma Berenices). Less than a degree to the southeast is M99. From here you can move 2° southeast to find two of the brightest galaxies in the cluster, at 9th magnitude, M84 and M86. They are both elliptical galaxies, and you should be able to see that M86 is larger and more elongated.

Continuing in the same direction you come to M87, a giant elliptical that is at the heart of the cluster. From there you can find M89 just over a degree to the east. Center on this elliptical, and with a low power you should find a more noticeable galaxy only 40 arc minutes to the northeast. This is M90, the first spiral galaxy along this route, and you will see that it looks more concentrated. Return to M89 and follow a route southeast to find M58, M59 and M60, a spiral and two ellipticals.

There is a second chain of galaxies farther south. To find these, return to M87 and move 4° south to the elliptical galaxy M49, then hop from there to the other galaxies in the chain, NGCs 4526, 4535, and 4365. There is one other bright galaxy in Virgo worth locating, and that is M104, the Sombrero Hat. To find it, locate Spica and then move 11° west. It looks like an oval with a strangely flat edge, resembling the wide-brimmed Mexican hat that gives it its name. Photographs show it as an almost edge-on spiral with a dark dust lane, which is what produces the straight edge. It is one of the easiest galaxies in Virgo as it is somewhat closer and probably not a true member of the cluster.

Object	Type	Mv	Magnification	Distance
M49	Galaxy	8.4	×75	60 million light years
M58	Galaxy	9.7	×75	60 million light years
M59	Galaxy	9.6	×75	60 million light years
M60	Galaxy	8.8	×75	60 million light years
M84	Galaxy	9.1	×75	60 million light years
M86	Galaxy	8.9	×75	60 million light years
M87	Galaxy	8.6	×75	60 million light years
M89	Galaxy	9.8	×75	60 million light years
M90	Galaxy	9.5	×75	60 million light years
M104	Galaxy	8.0	×75	30 million light years

► *The Hubble Space Telescope image of M104 reveals the dust lane across the center of the galaxy.*

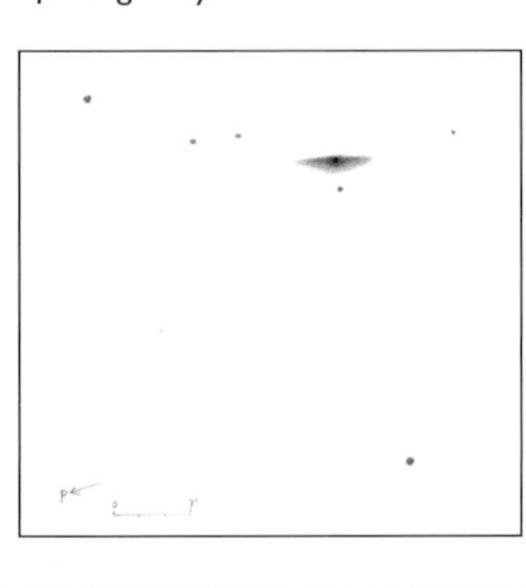

◄ *Through a 450 mm Dobsonian, the Sombrero Hat galaxy shows a distinct straight edge on one side.*

Coma Berenices

None of the stars of Coma is much brighter than 4th magnitude, but the constellation is striking. It contains a notable star cluster, looking to the naked eye much like other clusters do in a telescope. It is best seen in binoculars, and most telescopes are of less use because the cluster is so large, at about 5° across. The group is about 250 light years away and is simply known as the Coma Star Cluster. There is another Coma Cluster, of galaxies, but these are all 11th magnitude or fainter.

The stars of the Coma Star Cluster are meant to represent the hair of the Egyptian Queen Berenice, but the name itself is not ancient and dates from the 16th century.

There are numerous galaxies within Coma, some of which at the southern end of the constellation are part of the great Virgo Cluster. Like the other Virgo galaxies, if you do not have a GoTo telescope they are best located by starting at Denebola in Leo, moving east to 6 Comae,

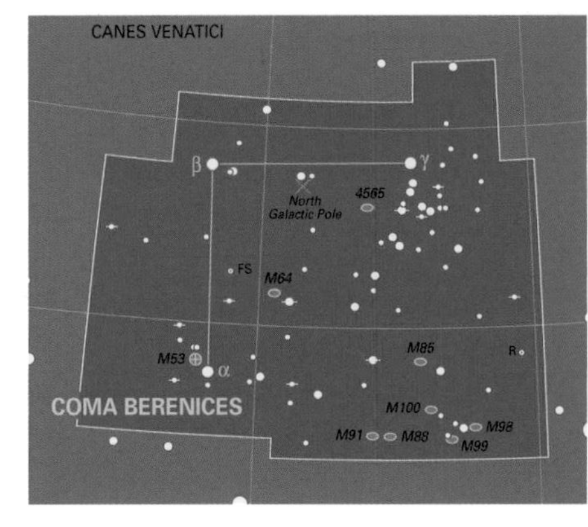

and moving from one to the other from there. Unlike many of the galaxies in the main part of the Virgo Cluster, those in Coma are mostly spirals rather than ellipticals, M85 being an exception.

M64, the Black Eye Galaxy

Coma has some galaxies that are closer than the Virgo Cluster, of which the brightest is M64. It is conveniently located near to the star 35 Comae, which you can find by star-hopping from Alpha Comae. The galaxy, which is about 19 million light years away compared with the 60 million of most of the Virgo Cluster, has a notable dark lane that supposedly lends it an eye-like appearance. However, this lane is not easily visible with small telescopes or under poor conditions, and it often looks like an oval smudge.

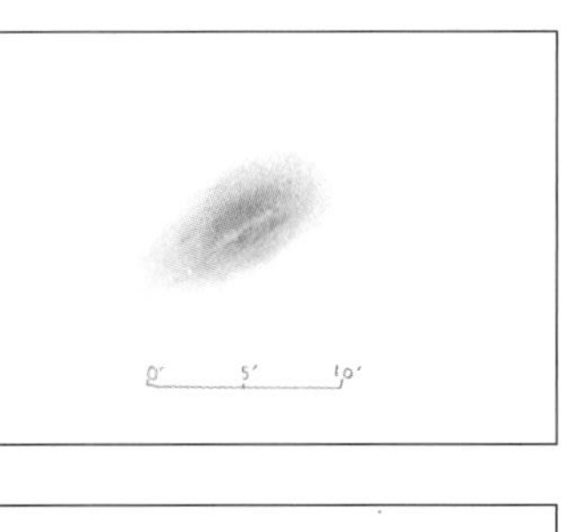

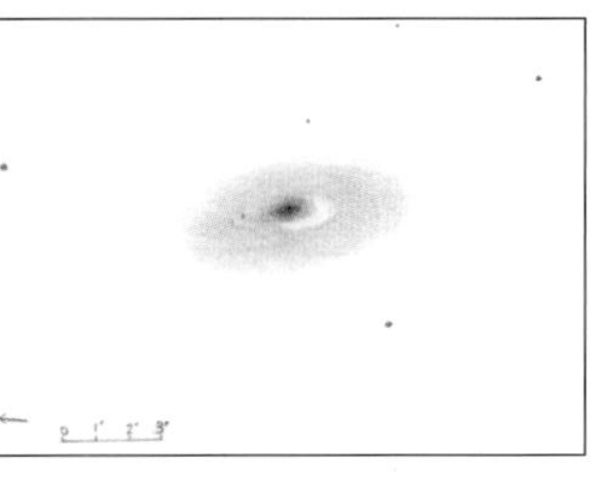

◀ *Through a 75 mm refractor (top), M64 shows a darker spot within its body. With a larger instrument (bottom), in this case a 450 mm, the galaxy takes on a more eye-like appearance.*

Galaxy NGC 4565

Edge-on spiral galaxies are popular with deep sky observers because they have a greater visual appeal than the featureless ellipticals and some of the blander spirals. This is one of the best and most popular, and even a small telescope will reveal it as a spindle with a central bulge. A medium to large telescope in a good sky will show a central dust lane as well. The object is 10th magnitude but is easier to see than some of the Messier objects in the Virgo Cluster because of its more condensed nature. It is easily found from the stars of the Coma Star Cluster.

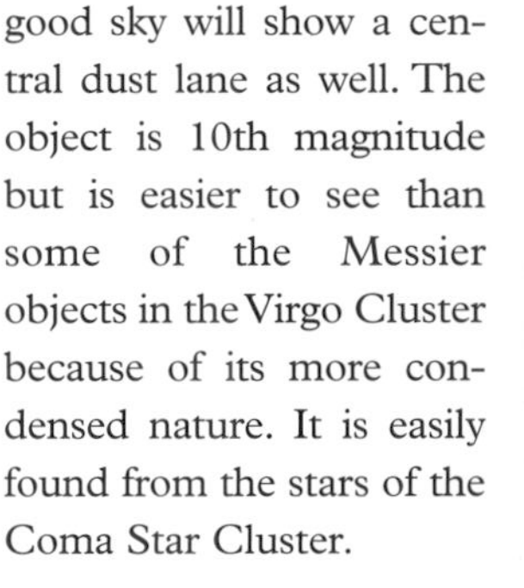

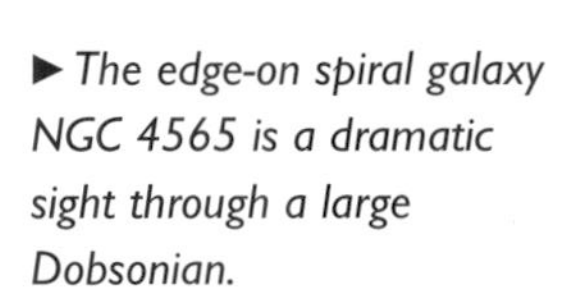

▶ *The edge-on spiral galaxy NGC 4565 is a dramatic sight through a large Dobsonian.*

Object	Type	Mv	Magnification	Distance
M64	Galaxy	8.5	×75	19 million light years
M99	Galaxy	9.9	×75	60 million light years
NGC 4565	Galaxy	9.6	×75	40 million light years

Leo, Leo Minor, and Sextans

One of the most ancient constellations, Leo occupies a central place in the evening skies around March, April, and May. Though just north of the celestial equator it can be viewed the world over, but only from the northern hemisphere does its actual resemblance to a crouching lion become obvious. Its most prominent feature is the reversed question mark of stars, which is widely referred to as the Sickle from its shape. This marks the lion's mane, and the rest of its body is marked by three other stars with Denebola as its tail. The word "deneb" crops up in a few star names, as it means "tail." Leo Minor, to its north, and Sextans, to its south, are comparatively recent inventions.

Being well away from the Milky Way, Leo is devoid of nebulae and clusters within our own Galaxy, but in their place we can see beyond to other galaxies. There are a few relatively bright galaxies within its boundaries. Leo is a zodiacal constellation, so the planets may be found there, and from time to time the Moon or a planet may be close to or even pass in front of Leo's brightest star, Regulus. A series of conjunctions, or close approaches, between Regulus and Jupiter in 2 BC is one theory behind the story of the Star of Bethlehem.

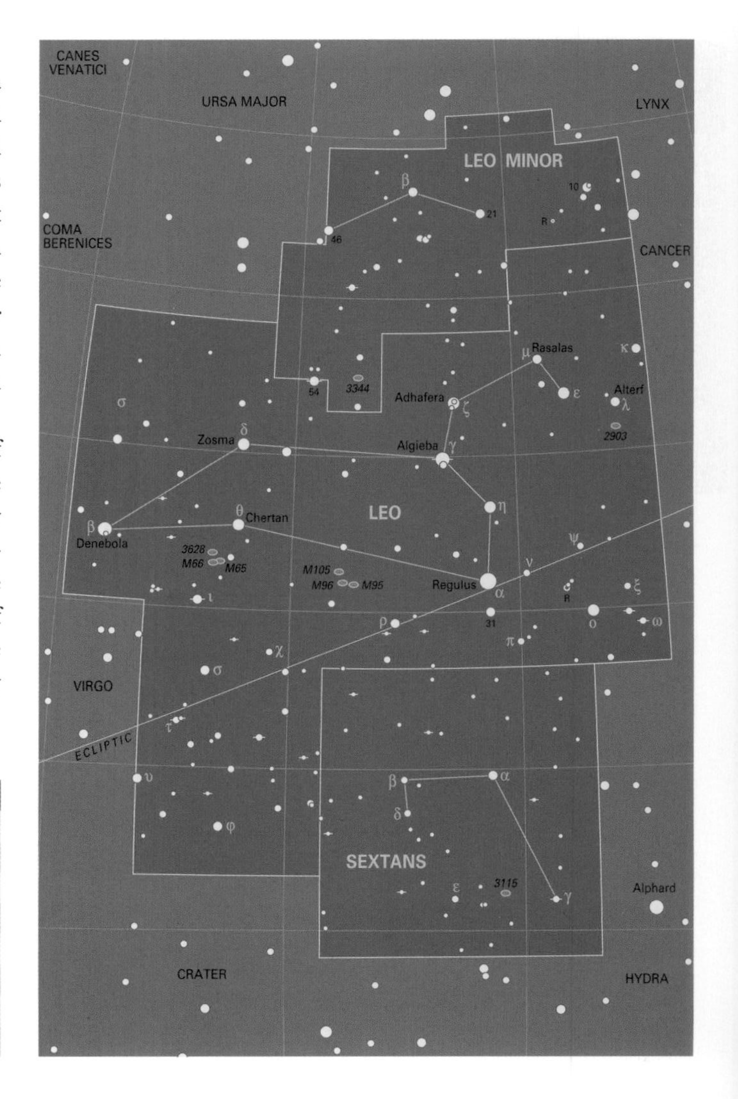

Object	Type	Mv	Magnification	Distance
Gamma Leonis	Double star	2.0	×200	126 light years
M65	Galaxy	9.3	×75	35 million light years
M66	Galaxy	8.9	×75	35 million light years
NGC 3628	Galaxy	9.5	×75	35 million light years
M95	Galaxy	9.7	×75	38 million light years
M96	Galaxy	9.2	×75	38 million light years
M105	Galaxy	9.3	×100	38 million light years
NGC 3115	Galaxy	9.2	×75	30 million light years

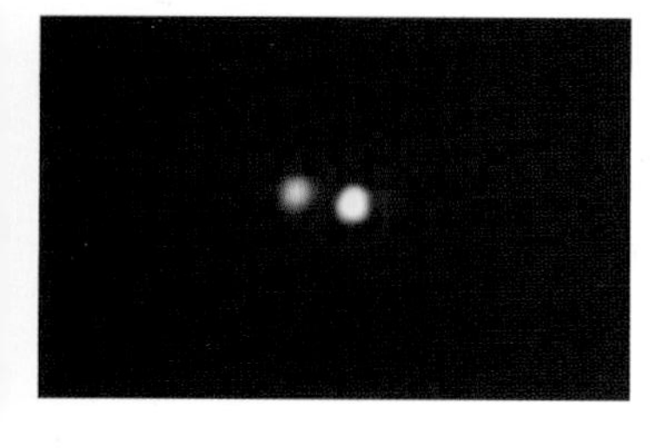

◀ *Double star Gamma Leonis (Algieba) presents two golden-yellow stars close together.*

▶ *Barred spiral galaxy M95, photographed with a 0.9 m telescope at Kitt Peak.*

Double star Gamma Leonis (Algieba)

Many doubles are famed for their beauty because they have contrasting colors, or because they are widely separated, but Algieba's stars are quite close, at just over 4 arc seconds, and both are yellowish in color. Nevertheless, their brilliance and closeness makes them a very popular pair, and because they are so close slight differences in their colors are often seen. The individual stars are types K and G, so some color difference is to be expected. A fairly high magnification is needed for the best views of this double.

Galaxies M65 and M66

A number of galaxies are within the reach of binoculars on a good night, but they are usually small and unremarkable. It always helps to have a pair of them, because this provides confirmation that you have seen a genuine tiny misty object rather than just maybe a faint star. These two galaxies in Leo are easily found by first locating 3rd-magnitude Theta Leonis, Chertan. About $2\frac{1}{2}°$ to its south is a vertical line of three 5th- and 6th-magnitude stars. M66 is at the apex of an equilateral triangle with these stars, with M65 nearby. Both are noticeably elongated spiral galaxies, and a small telescope shows them easily, with some detail visible using larger apertures. Having found these galaxies, you may spot the more elongated spiral NGC 3628 to the north.

◀ *The trio of galaxies M65 (lower right), M66 (lower left), and NGC 3628 (top).*

Galaxies M95, M96, and M105

Having found M65 and M66, you can either move exactly 8° to the west to find M105 in the same field of view, or you can move 9° east of Regulus to find this group of galaxies. They are slightly fainter than M65 and M66, so binoculars may not show them, but they can be easily located using a low power on a small telescope. M95 is a good example of a barred spiral galaxy, though with a small telescope only the nucleus is visible and the bar requires a larger aperture. M96 is a spiral and M105 an elliptical galaxy.

Lenticular galaxy NGC 3115

The brightest example of a lenticular galaxy, this appears as a smooth spindle in a telescope, and indeed it has the nickname of the Spindle Galaxy. It is relatively easy to find by moving just over 3° eastward and very slightly northward from 5th-magnitude Gamma Sextantis. It has a fairly bright center, so it is easily observable with medium-sized telescopes. This is the only bright object of any note in Sextans.

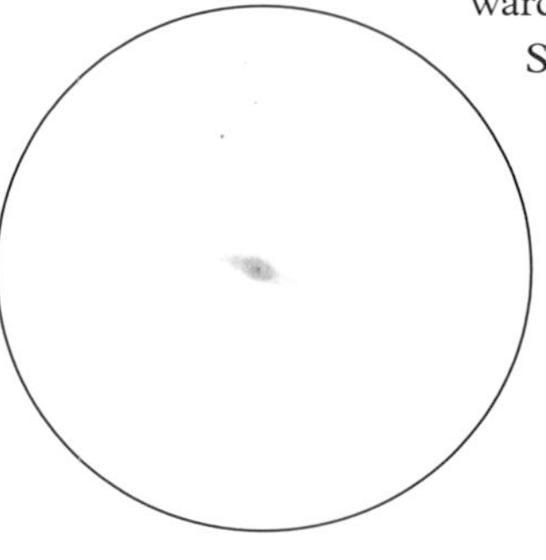

◀ *A drawing of the lenticular galaxy NGC 3115 made with a 222 mm reflector.*

Cancer

Everyone has heard of the constellation of Cancer as it is in the zodiac, but this is no guarantee that it has bright stars. In fact, only two of its stars are brighter than 4th magnitude. Nevertheless, Cancer is an ancient constellation containing some interesting objects, which of course may include the Moon and planets. It is a northern constellation but is easily visible from southern-hemisphere locations.

Star cluster M44, the Beehive or Praesepe

To the naked eye, M44 appears as a little misty patch just to one side of the line between the stars Gamma and Delta Cancri. It is visible in clear skies even if there is a certain amount of moonlight. Binoculars reveal it as a large cluster of stars from 6th magnitude downward, and in a telescope at the lowest power

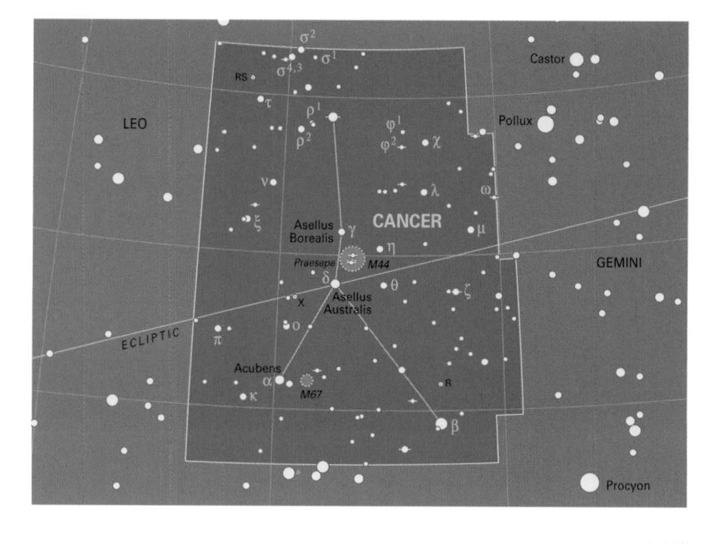

the eyepiece is full of stars. In longer-focal-length instruments, however, not all of the cluster will fit inside the field of view. The cluster is about 590 light years away, putting it somewhat more distant than the Pleiades. The word *praesepe* is Latin for "manger," referring not to the manger of Christ's birth but to the stars Gamma and Delta, known as the *aselli* or donkeys.

◄ *The open cluster M44, known as the Beehive or Praesepe, flanked by the two donkeys.*

Star cluster M67

Easily found by star-hopping from Alpha Cancri, this is a much more distant and hence smaller and fainter cluster than M44. A medium-sized telescope is needed to do it justice, as most of its stars are 10th magnitude or fainter. M67 is a particularly old cluster at around 4 billion years, compared with 730 million years for M44.

▲ *In contrast to M44, M67 is a challenge to binocular users as it is much smaller and fainter.*

Object	Type	Mv	Magnification	Distance
M44	Open cluster	3.7	×7	590 light years
M67	Open cluster	6.1	×25	2700 light years

GROUP 10

Andromeda

The region around Andromeda is a mythological cartoon strip. Andromeda, the daughter of Queen Cassiopeia, is chained to a rock as a sacrifice to the sea monster Cetus. In adjacent frames we see the hero Perseus who rescues her, and her worried mother Cassiopeia. The whole story remains in the sky to be told and retold around camp fires.

The four main stars of Andromeda are in a rough line, and hardly represent a beautiful maiden, but the constellation does include one of the sights of the sky, the Andromeda Galaxy. At one end of the line is Perseus – in fact, you can reach Alpha Persei by continuing the line of stars to the east. At the other end is the Square of Pegasus, and the star Alpha Andromedae doubles up as one corner of the Square. Andromeda is a northern constellation, but most of its stars can be seen from the southern hemisphere, though very low down for the more southerly observers.

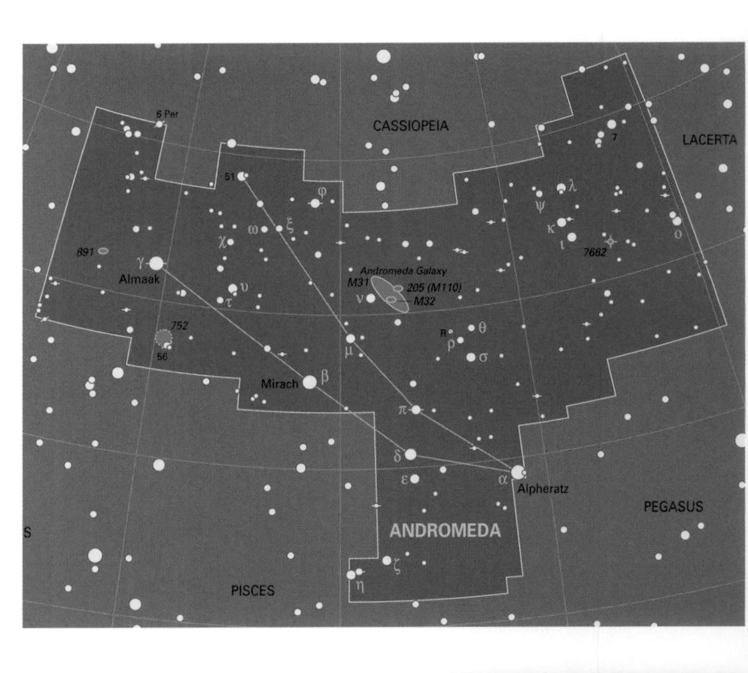

Double star Gamma Andromedae (Almaak)

▲ *The closeness and contrasting colors of its two stars make Gamma Andromedae particularly beautiful.*

Contrasting colors are always popular in double stars, and Gamma Andromedae is a good example. Its stars are yellow and blue, separated by just under 10 arc seconds, so even a small telescope will split them. The 6th-magnitude secondary star is itself a double, though the two stars are very close together.

M31, Andromeda Galaxy, M32 and NGC 205 (M110)

It comes as a surprise to many people that the Andromeda Galaxy is not only easy to find, but also visible with the naked eye even in average skies. With binoculars it can be viewed even from city centers on clear nights. Although it appears as only an oval smudge under such circumstances, it is something just to be able to observe the heart of another galaxy over 2.5 million light years away armed with nothing more than binoculars or the naked eye. The Andromeda Galaxy is generally regarded as the most distant object that can be seen with the naked eye, though there are always people who can see more distant objects under the right conditions.

To locate it, find the line of stars Alpha, Delta, Beta, Gamma. From Alpha, go two

▲ *Locating the Andromeda Galaxy. The bright star at lower left of center is Beta And; from there move up past Mu and Nu to find M31 itself.*

▲ *A close-up of galaxy M32, taken by Peter Shah from Meifod, Wales. For an amateur drawing of M31 and M32, see page 81.*

stars along to Beta and from there take a right angle to the north and count along two fainter stars, Mu and Nu. The Andromeda Galaxy is alongside Nu Andromedae. If you don't see it by direct vision, look away from it slightly and you should pick up a misty ellipse. In light-polluted skies you are seeing just the nucleus of the galaxy, but from a good country site and with good eyesight you can trace it much farther out – its outline on the map shows that it covers more than 3° of sky.

In a telescope, the smudge becomes larger and more extensive, and you should be able to pick up M31's two companion galaxies, M32 and NGC 205. M32 is a circular elliptical galaxy; on photographs it appears on the edge of M31, but visually it is usually separate. NGC 205, now generally added to Messier's catalog as M110 on the grounds that he did actually observe the object, is an ellipse on the other side of M31 from M32. If it were not for M31, both these companion galaxies would still be among the brightest in the sky, but we tend to relegate them to minor objects because of their proximity to the great galaxy itself.

Large telescopes will show a surprising amount of detail in M31. The dark lane shown on photographs is visible in good skies with medium apertures, but individual star clouds and globular clusters can be made out under good conditions.

Open cluster NGC 752

Some clusters, like this one, are comparatively faint but quite rich in stars. This scattering of stars of 8th to 10th magnitude is particularly attractive in a telescope, though it is also visible in binoculars. The unrelated star 56 Andromedae is a good guide to its location.

◀ *NGC 752 is the scattering of stars below the center. The double star just below it is 56 And.*

Planetary nebula NGC 7662, the Blue Snowball

With a nickname like "the Blue Snowball," the popularity of this bright planetary nebula is assured. Despite the lack of nearby bright stars it is fairly easy to find without a Go To telescope, as it lies just under 5° to the east of the pair of stars Omicron Andromedae and 2 Andromedae. With these in the field of view of the telescope, either sweep eastward or leave the telescope stationary for 23 minutes and the planetary will drift into the center of the field of view. It lies near an 8th-magnitude star, and is of similar brightness but noticeably larger, even with a low power. The blue color is not as obvious to every observer as its name would imply. Higher magnifications will help to reveal some internal detail, particularly with larger instruments.

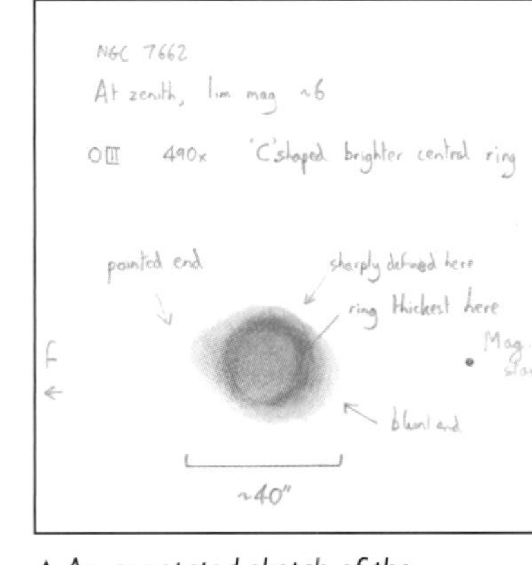

▲ *An annotated sketch of the Blue Snowball, made with a 250 mm reflector.*

Galaxy NGC 891

Andromeda is host to a few galaxies in addition to the M31 group, and NGC 891 is the brightest example. It requires a medium aperture, but is quite easily found by sweeping $3\frac{1}{2}°$ eastward from Gamma Andromedae. It lies midway between that star and the cluster M34 in Perseus. It is an edge-on spiral, and the dark lane that bisects it can be seen under good conditions.

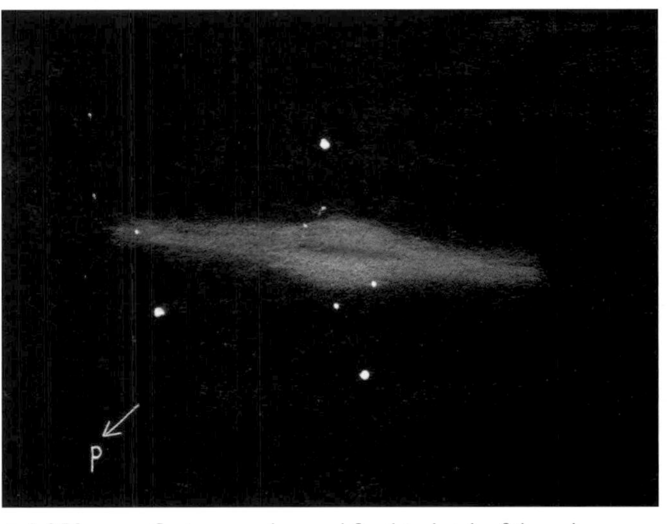

▲ *A 250 mm reflector was also used for this sketch of the galaxy NGC 891, with powers of 113 and 184.*

Object	Type	Mv	Magnification	Distance
Gamma And	Double star	2.1	×200	355 light years
M31	Galaxy	3.4	×25	2.5 million light years
M32	Galaxy	8.1	×75	2.5 million light years
M110	Galaxy	8.5	×75	2.5 million light years
NGC 752	Open cluster	5.7	×25	1300 light years
NGC 7662	Planetary nebula	8.3	×250	2200 light years
NGC 891	Galaxy	10.0	×75	10 million light years

Cassiopeia

To those viewing in the northern hemisphere the W-shape of the main stars of Cassiopeia (or M-shape, depending on which way up it is) is a familiar sight – for it is always in the sky for most observers. As usual, its star pattern has very little to do with its representation of Queen Cassiopeia on her throne. It is in a rich area of the northern Milky Way, and although it lacks the very brightest of splendors, there are many fainter deep sky objects within it. It was also the location of one of the few supernovae to be seen in our Galaxy in the past millennium, in 1572, though no visible trace of this event remains.

The star Gamma Cassiopeiae, the middle star of the W, is an erratic variable star. For most of the time it remains at about magnitude 2.2, but it can fade to magnitude 3 and has been known to brighten to 1.6. A blue giant star, it from time to time throws off shells of material and could undergo another shell episode at any time. Unusually for such a prominent star, it has no accepted popular name and is usually known by the abbreviation Gamma Cas.

Long-exposure photographs on fast film show several pink wisps of nebulosity in Cassiopeia, notably IC 1805 and IC 1848.

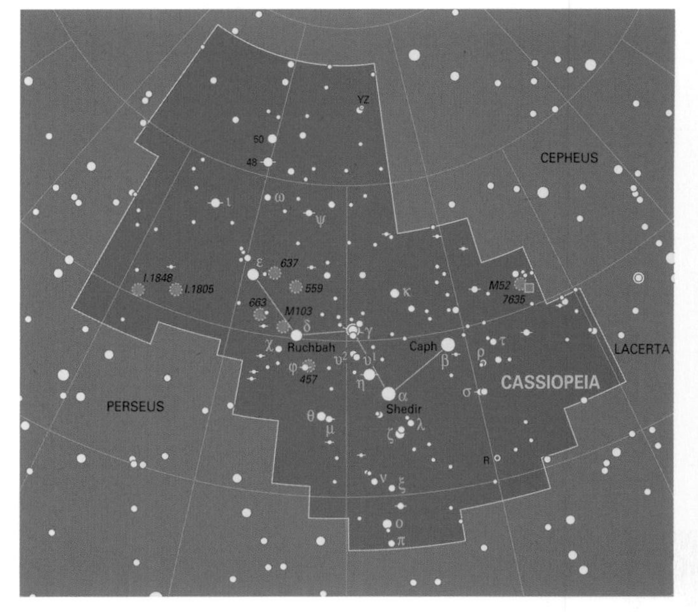

Open cluster M52

Follow a line from Alpha through Beta Cas an equal distance and you come to the open cluster M52. It is visible with binoculars as a misty blur, and a small telescope will show numerous stars of about 10th magnitude. It is quite compact and distant, and requires more magnification than many other clusters. Nearby lies a nebula, NGC 7635, known as the Bubble Nebula. In medium and large telescopes, under good conditions, it is just visible as a crescent-shaped nebula surrounding a star. Photographs reveal a curious circular bubble which results from hot gas expelled from a central star.

▲ *A small telescope will show cluster M52 (top left), but a larger telescope or a long-exposure photograph is needed to reveal the Bubble Nebula at bottom right.*

NGC 457, Owl Cluster

Many clusters have nicknames, but this one has at least two aliases – the Owl Cluster and the ET Cluster. At one edge of it is 5th-magnitude Phi Cas, just south of Delta Cas. Like the Jewel Box in Crux, NGC 457 is a cluster to which binoculars do not really do justice because the bright stars overpower the underlying fainter stars. The alternative names derive from two long streams of stars at an angle to the main cluster, which some see as the wings of an owl and others as the arms of Spielberg's famous extraterrestrial. In a telescope the arms or wings become very obvious, and Phi Cas and a neighboring star look like the huge eyes of an owl or of ET.

Object	Type	Mv	Magnification	Distance
M52	Open cluster	7.3	×50	5000 light years
NGC 7635	Nebula	–	Photographic	7100 light years
NGC 457	Open cluster	6.4	×25	2400 light years
M103	Open cluster	7.4	×50	8500 light years
NGC 663	Open cluster	7.1	×25	2000 light years

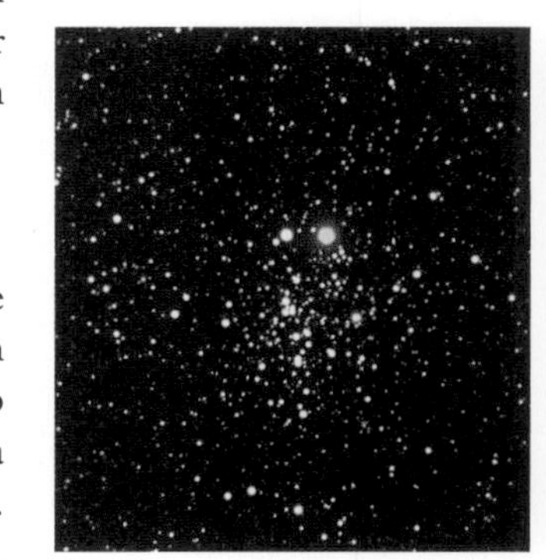

▲ *The two bright stars in NGC 457 have the appearance of huge eyes, an effect that is even more vivid in a telescope than on a photograph.*

Open cluster M103

Smaller and less impressive than NGC 457, M103 can be found by returning to Delta Cas and moving a degree east and slightly north. Although visible in binoculars, a higher magnification is a great help. M103 is noticeably fan-shaped. Beyond it is the cluster NGC 663, which is more obvious in binoculars.

▲ *M103 requires a telescope for the best views and is not to be confused with the nearby and larger NGC 663.*

Triangulum

This is a constellation that no one can argue does not resemble its namesake – a triangle. It dates from Greek times, but consists of stars of only 3rd and 4th magnitude. Its modern claim to fame is that it is home to the galaxy M33, the Pinwheel Galaxy. This is a member of our Local Group of galaxies, of which the other leading members are our own Galaxy and M31. The Pinwheel is about 3 million light years away, so is slightly more distant than M31 and its companion galaxies.

Galaxy M33, the Pinwheel Galaxy

For those few observers able to see this galaxy with the naked eye, it is the most distant object visible. But the majority of us need binoculars. If you are used to galaxies in binoculars being tiny blurs among the stars, you may overlook M33 as it is very large. Once you are aware of this, however, it is not too hard to find in reasonably dark skies. The map shows its location not far from Alpha Tri, and it lies about a binocular field away from that star.

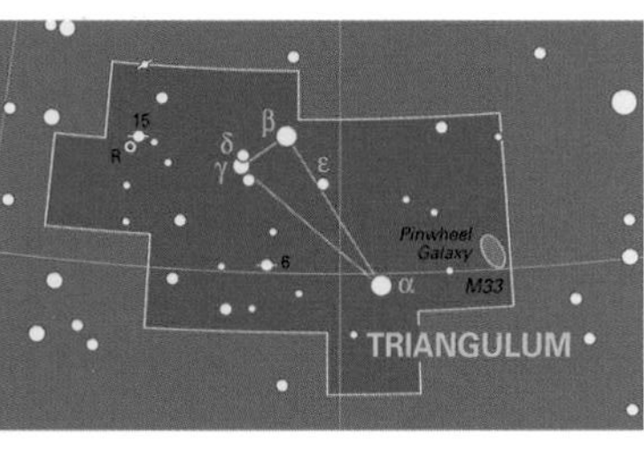

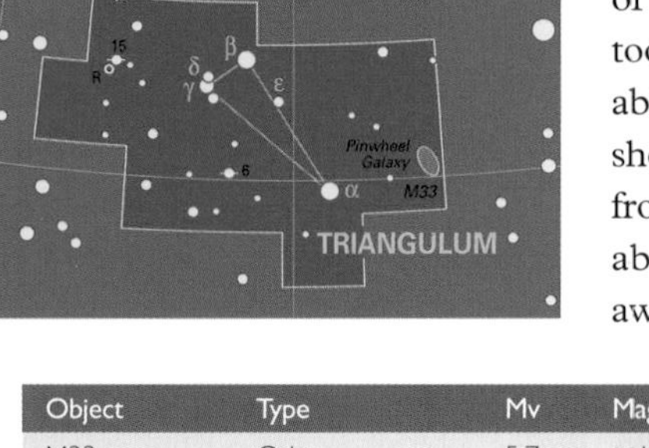

Object	Type	Mv	Magnification	Distance
M33	Galaxy	5.7	×10	3 million light years

▲ *The galaxy M33 photographed with the Mayall 4 m telescope at Kitt Peak.*

Though the nucleus appears to be surrounded by a fainter area, the spiral structure that is seen in photographs is not obvious. The total brightness is magnitude 5.3, but this is spread over an area twice the diameter of the Full Moon so the surface brightness is low.

▲ *This telephoto lens shot of the region around M33 has a field of view of 5.5° and shows stars down to magnitude 11. It closely resembles the galaxy's appearance in binoculars.*

GROUP 11

Hydra

The largest constellation of all, this sea snake wriggles over a great swathe of the midsouthern sky, squeezing between several other constellations on its way. Its head is in the northern hemisphere, and while all of it can be seen by most northern observers, some parts of it are so low that they are effectively unobservable. Hydra covers nearly seven hours of right ascension, but for all its size it is a barren area of sky and has fewer than its fair share of deep sky objects.

The Head of Hydra is its most recognizable pattern, a group of six 3rd- and 4th-magnitude stars lurking below Cancer. The brightest star, Alpha Hydrae, is in an isolated position to the southeast of the head, and its name, Alphard, does indeed mean "Solitary One" as there are no other nearby bright stars.

Below Spica, one of the wriggles in Hydra's body is marked by the star R Hydrae, which is often invisible as it is a Mira-type star (see page 100), varying between magnitudes 4 and 10 in about 390 days.

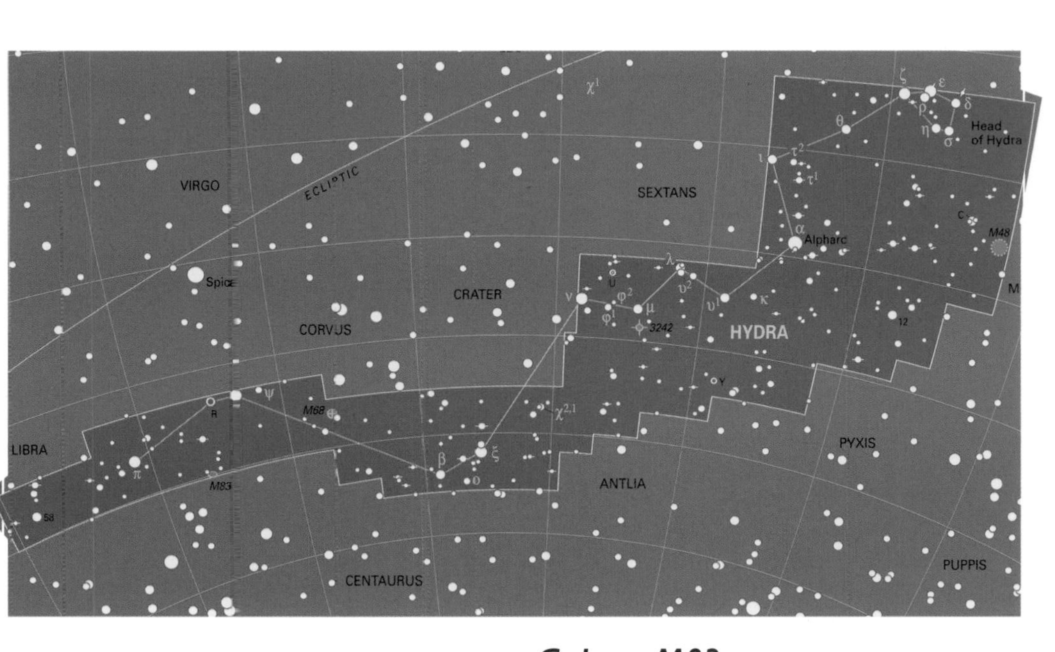

Object	Type	Mv	Magnification	Distance
M83	Galaxy	7.6	×25	15 million light years
NGC3242	Planetary nebula	7.8	×100	2500 light years

Galaxy M83

A classic face-on spiral galaxy, this is a very difficult object for northerly observers, but when it is high up and in dark skies it is a splendid sight, being one of the few galaxies whose spiral arms are visible with amateur telescopes. The lack of nearby bright stars makes M83 a little hard to find. It is at one apex of a large triangle with Pi and Gamma Hydrae, with a short line of 5th- and 6th-magnitude stars to its west, and another star to its east. These stars should be visible in the finder to help you locate the galaxy itself.

If the galaxy is low down and hard to find in the murk, you are likely to see only its nucleus. But the better the skies, and the higher it is, the more detail you should see, even with medium apertures. Wait until it is as high in the sky as possible for the best views.

▲ *M83 is sometimes referred to as the "Southern Pinwheel" for its resemblance to M33. Though smaller and fainter than that galaxy, it is easy to see even with binoculars under the right conditions.*

Planetary nebula NGC 3242, the Ghost of Jupiter

As if to prove that not every object in Hydra is out of reach of northern observers, the Ghost of Jupiter planetary is easily visible even in small instruments. It gets its name from its similarity to the planet itself – in terms of size and shape but not brightness or color, hence the ghostly appearance. The object is 8th magnitude and is large enough to be found with little trouble, once you have the right part of sky. Find Alphard and then locate 3rd-magnitude Mu Hydrae. From there, the Ghost of Jupiter is just 2° south and a little to the west, and many observers notice a blue color. Larger telescopes reveal a fainter outer shell.

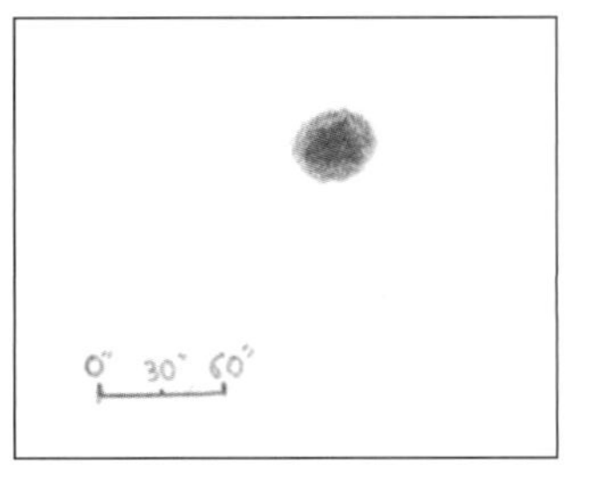

▶ *The Ghost of Jupiter, as seen through a 450 mm Dobsonian, though smaller apertures will easily show it.*

Centaurus

Despite the fact that it lies below declination –30° and is therefore largely unobservable from Europe, Centaurus is an ancient constellation. Over the past 2500 years, a slow movement of Earth's axis, known as precession, has resulted in its stars slipping southward, so Centaurus is now regarded as a southern constellation. However, almost all of it is visible from the most southerly extremities of the United States.

Centaurus is a difficult constellation to get to grips with because much of it consists of scattered 2nd- and 3rd-magnitude stars with no clear pattern. But its leading stars are unmistakable – a brilliant pair usually known as Alpha and Beta Centauri rather than by their names of Rigil Kent and Hadar. They are known as the Pointers because they indicate the nearby Southern Cross – not that this famous group needs much additional identification.

From Alpha and Beta a couple of straggling lines of stars spread northward and westward. Though the northern extremities of the constellation are rather barren, the southerly end of Centaurus is in a rich part of the Milky Way which contains a number of clusters, though none is particularly bright or rich.

◀ *This deep view of the sky around Alpha Centauri (overexposed at top left) shows the vast difference in brightness between it and Proxima, marked by the arrow.*

▶ *A close-up of Alpha Centauri shows that the two stars are slightly different in brightness and color. They are separated by 14 arc seconds.*

Alpha Centauri

This is the nearest bright star to the Sun, a mere 4.38 light years away, and the third-brightest star in the sky. But unlike the Sun, it is a multiple star. Even a small telescope shows that it is a double, with one star somewhat brighter than the other. The brighter of the two is a type G star like the Sun, and is very similar in true brightness. Its companion, however, is a fainter type K star that orbits it every 80 years. Despite science fiction stories, no planets have been detected around the star. If there were any, the inhabitants would see a night sky very similar to ours except that there would be an extra star at 1st magnitude in Cassiopeia – our Sun.

There is a third member of the system, Proxima Centauri, which is slightly nearer to Earth, at 4.28 light years away, and is therefore the closest star to the Sun. However, it is a dim red dwarf and is only 11th magnitude, so it is not readily seen. Nor is it close to Alpha itself – it is 2° away, well outside the field of view of most telescopes.

Globular cluster Omega Centauri

Normally, a designation like Omega Centauri would be given to a star, but in this case the 4th-magnitude star that you can see in the sky looks distinctly more fuzzy than the others. You can use the stars of the Southern Cross to point to it. Turn binoculars or a telescope on this star and it is clearly a globular cluster, though unwary observers using low powers might think that they have discovered a new tail-less comet. It is not just any old globular, however, but the brightest in the sky. It is a must-see object for anyone who lives far enough south to observe it well – which means anywhere south of the Mediterranean, or the southern states of the United States. A small telescope will resolve some stars, and the larger the aperture the more amazing the view. In a medium or large telescope under good conditions the sight is generally described using superlatives such as "awesome." The cluster contains around a million stars, and it is one of the nearest globular clusters.

◀ *Omega Centauri's resemblance to a comet is shown in this 1986 photo of Halley's Comet, the turquoise object above center, and Omega itself, at bottom right. NGC 5128 is visible to the right of the comet.*

Object	Type	Mv	Magnification	Distance
Alpha Cen	Double star	–0.3	×50	4.38 light years
Proxima Cen	Star	11.0	–	4.22 light years
Omega Cen	Globular cluster	3.7	×25	16,000 light years
NGC 3918	Planetary nebula	8.4	×100	3000 light years
NGC 5128	Galaxy	7.0	×75	12 million light years

NGC 3918, the Blue Planetary

Several planetary nebulae are bright enough to appear blue, but many observers comment on the blueness of this one. Find it by using Crux to indicate a little triangle of 5th- and 6th-magnitude stars. It is fairly small so a considerable magnification is needed to show a fuzzy disk with little internal detail.

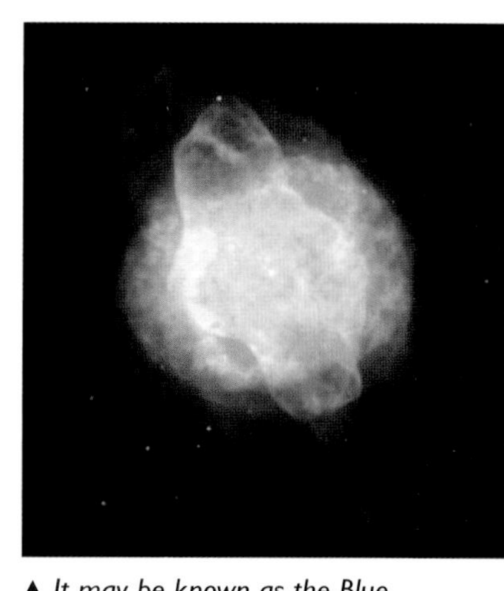

▲ *It may be known as the Blue Planetary, but this Hubble view of NGC 3918 emphasises different colors.*

Galaxy NGC 5128

If you are observing Omega Centauri, move just under 5° to the north and you will spot this peculiar galaxy, which, at 7th magnitude, is one of the brightest in the sky. It is noticeable in binoculars, often in the same field of view as Omega but very much smaller, appearing as a small fuzzy glow. A small telescope shows it as a circular patch of light, but with a medium to large telescope you can see a dark lane bisecting the center. One nickname for this object is "the Hamburger Galaxy."

This is one of the most thoroughly observed galaxies in the sky as far as professional astronomers are concerned. The dark lane seems to be the remains of a spiral galaxy that has collided with the main elliptical galaxy, and it is a major source of radio waves, known as Centaurus A.

▲ *Giant galaxy NGC 5128 (Centaurus A), photographed by the Very Large Telescope in Chile.*

Crux and Musca

There can be few people who have not heard of the Southern Cross. The Latin version of the name is simply Crux, meaning the Cross. As we have become used to these stars on the flags of several nations, it may seem odd that the group has not always been a separate constellation. Until the 17th century, however, the stars were regarded as part of Centaurus, though they have not been visible from Europe since ancient times because of precession. It was when European navigators first saw the pattern that it became regarded as a cross, and it is now generally taken as a sign that you have arrived in the southern hemisphere, despite being visible from as far north as the southern tip of Florida in the United States.

The Southern Cross can be used as a quick means of indicating the south celestial pole. Northern observers have the luxury of a bright star, Polaris, within a degree of the north celestial pole. But there is no such bright star in the southern hemisphere, so observers must make more of a guess. The long axis of the cross points roughly north–south, but by itself this is not enough. Bisecting the line between Alpha and Beta

Object	Type	Mv	Magnification	Distance
NGC 4755	Open cluster	4.2	×50	7600 light years

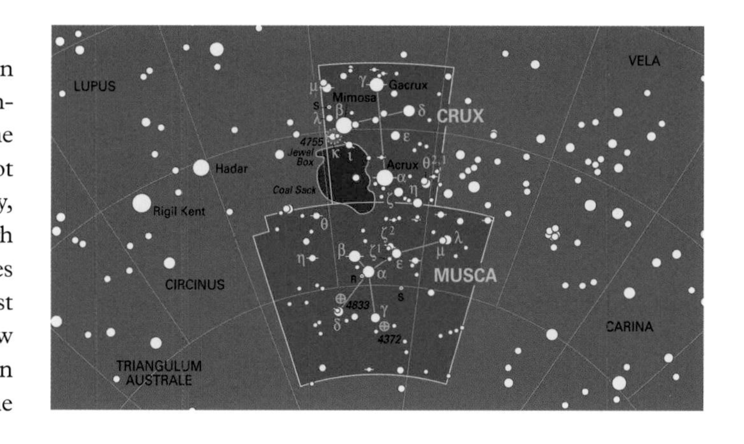

◀ *How to use the Southern Cross and the Pointers to find the south celestial pole.*

Centauri provides another line, and the two intersect about 3° from the true pole.

Crux is the smallest constellation of all, but its location in the Milky Way provides it with several interesting objects. Alpha Crucis is a double star that can be resolved using small telescopes, the separation of the two stars being 4.4 arc seconds. Gamma is a red giant star, appearing noticeably orange compared with the other three bright stars of the cross, which are all type B.

Musca, the Fly, is one of those constellations devised in the 16th century to fill in an otherwise unclaimed area of sky. The globular cluster NGC 4833 is an easy object to find just north of Delta Muscae, though it requires a medium aperture to resolve some stars.

◀ *A photograph of the Jewel Box Cluster with a large telescope shows the contrasting colors of the bright stars.*

NGC 4755, the Jewel Box or Kappa Crucis Cluster

Binocular observers, having heard that this is the most beautiful cluster in the sky, may wonder if they are looking at the right spot, because all they see is a knot of stars with none of the haze that signifies a rich cluster. The object is 1° from Beta Crucis, and is also known as Kappa Crucis after the cluster's brightest star, which is of 5th magnitude. With a telescope and a moderate power, however, the object becomes transformed. The stars are mostly bluish, with the exception of an 8th-magnitude red giant near the center. The color contrast has led to the cluster's popular name.

The Coalsack Nebula

If the sky is sufficiently clear that the Milky Way is visible, it becomes obvious that to the east of the Cross is a dark region, apparently devoid of stars. This is a dark nebula of the same type as those of the Cygnus Rift (see page 13). Although there is at least one star of naked-eye visibility within its borders, the effect is nevertheless striking. It is about 600 light years away.

▲ *The Coalsack is a dark cloud in an otherwise bright part of the Milky Way, close to the Southern Cross.*

The Magellanic Clouds

Many keen stargazers spend a lot time trying to locate distant galaxies, yet there are two galaxies that anyone far enough south can see at virtually any time – the Large Magellanic Cloud (LMC) and the Small Magellanic Cloud (SMC). Satellite galaxies to our own Galaxy, they are respectively a mere 160,000 and 190,000 light years away, compared with 2.5 million light years for the Andromeda Galaxy, M31. They are of the type known as irregular galaxies, meaning that have no regular form, unlike spiral or elliptical galaxies. The vast majority of galaxies are irregular, so these are probably what we should think of as a typical galaxy, rather than the spirals and ellipticals that attract so much of our attention. The LMC has a hint of organized structure, with an attempt at being a barred spiral.

These galaxies are named after the Portuguese navigator Ferdinand Magellan (1480?–1521), who voyaged south in 1519. They were certainly known to earlier European sailors, and were referred to as "Cape Clouds" after the Cape of Good Hope. Magellan's sailors used them as navigational aids. Though they lie within 20° of the south celestial pole, they can be seen at certain times of year, such as evenings from November to February, from Central America, West Africa, and southern Arabia and India. From much of the southern hemisphere they are visible all night. The LMC occupies part of the constellations of Dorado and Mensa, while the SMC is in Tucana.

They are often described as resembling fragments of the Milky Way that have become detached – they have a similar brightness to its brightest regions, and the LMC can be seen despite some light pollution or moonlight. The bright stars Canopus and Achernar act as rough guides – the LMC is about 15° south of Canopus, while the SMC is a similar distance south of Achernar. With the naked eye there is little detail visible, but binoculars hint at the amount of objects to be seen within.

Object	Type	Mv	Magnification	Distance
LMC	Galaxy	0.1	Naked eye	160,000 light years
NGC 2070	Nebula	8	×50	170,000 light years
SMC	Galaxy	2.3	Naked eye	190,000 light years
47 Tuc	Globular cluster	4.0	×25	13,400 light years

◄ A wide-field view of the southern skies shows Canopus (lower left), Achernar (upper right), the Large Magellanic Cloud (LMC, lower center), and the Small Magellanic Cloud (SMC, right).

Large Magellanic Cloud

The LMC has a diameter of 6° and is noticeably elongated along the line of the bar. A brighter patch within it, which to the naked eye resembles a hazy star just outside the main bar, becomes easily visible in binoculars. This is a massive diffuse nebula or H II region, known as the Tarantula Nebula. Its spidery extensions are visible in telescopes, particularly using O III filters. The Tarantula (NGC 2070) is huge: it is not very much smaller in the sky than the Orion Nebula, yet it is over 100 times more distant. By comparison, the Orion Nebula is insignificant, and would be virtually invisible if it were in the LMC. The Tarantula is a major star-forming region and a cluster of new stars, known as 30 Doradus, is visible in small telescopes at its center.

▲ The Large Magellanic Cloud with the Tarantula Nebula clearly visible to its upper left.

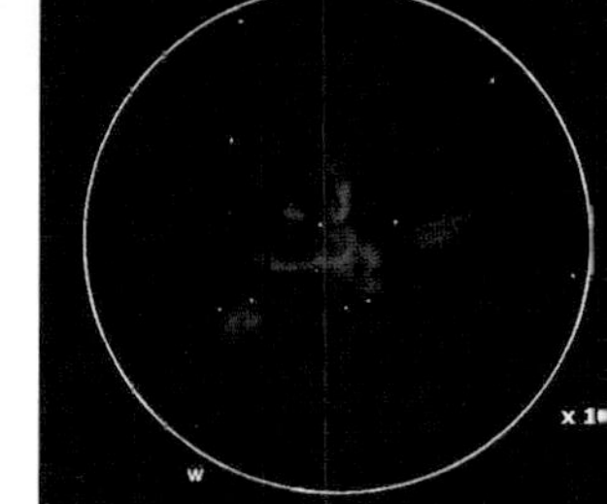

► A drawing of the Tarantula Nebula made using a 250 mm Dobsonian.

Numerous nebulae, star clusters, and globular clusters are to be seen within the LMC, with the brighter individual stars being visible in some of the open clusters with medium to large telescopes.

In 1987, a star appeared to the naked eye within the LMC, in addition to the few foreground stars within our own Galaxy that are visible within its area. This was a supernova, the brightest in terms of apparent brightness to have been witnessed in modern times. It rose to magnitude 2.8 before taking a year or so to fade below naked-eye visibility. It is referred to as SN 1987A, and remains the subject of intense study by professional astronomers.

The Small Magellanic Cloud

Smaller, fainter, and slightly more distant than the LMC, the SMC requires a darker sky to be seen with the naked eye. It resembles a teardrop or tadpole. The tail of the tadpole contains several H II regions, visible with medium to large telescopes with an O III filter. A round nebulous object within the SMC is the globular cluster NGC 362, which is actually much closer than the Cloud and just happens to be in the same line of sight. Were it separate from the SMC, it would be ranked as one of the brighter globulars in the sky.

Anyone glancing at the SMC will undoubtedly be drawn to its near neighbor in the sky, the spectacular globular cluster 47 Tucanae. As its star name rather than NGC number implies, it appears to the eye as a star of about 4th magnitude. It ranks second only to Omega Centauri in splendor, and with a telescope of more than 100 mm aperture you can start to resolve it into stars. Some observers prefer the appearance of 47 Tucanae to Omega Centauri because, being smaller, it fits better into the low-power field of view of a telescope. The cluster contains upward of half a million stars.

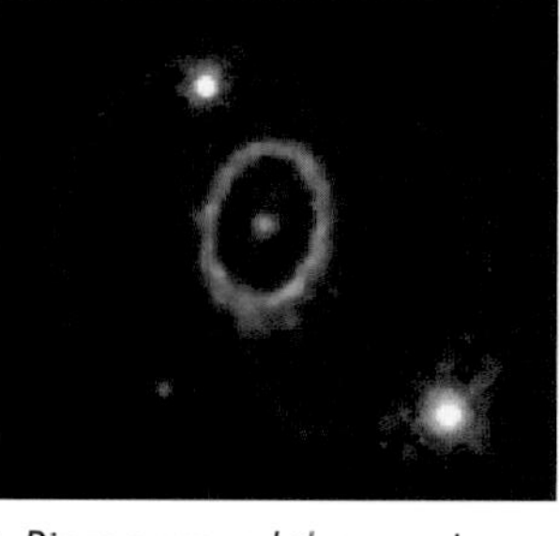

▲ Rings surround the remains of Supernova 1987A in the LMC, photographed with the Hubble Space Telescope in 1994.

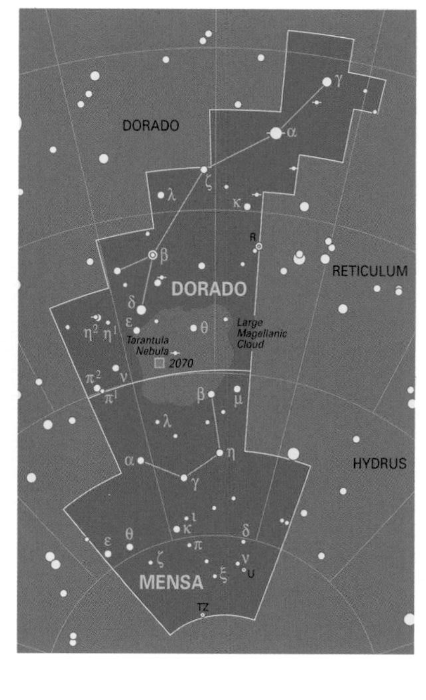

▼ Globular cluster 47 Tucanae, photographed with a 4 m telescope in Chile.

▲ The Small Magellanic Cloud and the globular cluster 47 Tucanae to its right.

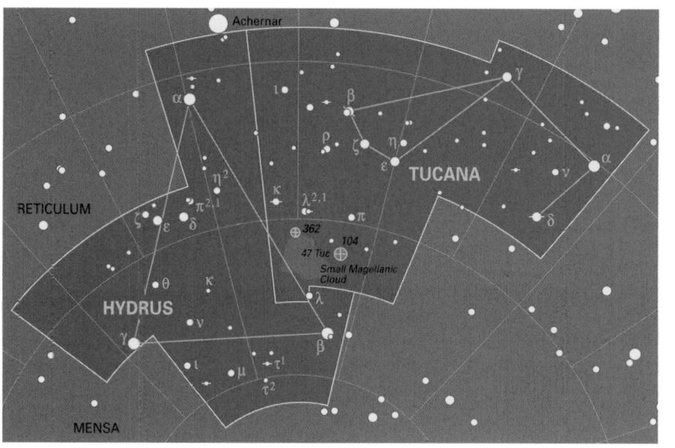

Planetary Data

In the data given below, the orbital period is the time taken for the planet to orbit the Sun – its year. The rotation period is the time it takes to spin on its axis – its day. The orbital eccentricity is the amount by which the shape of its orbit deviates from a circle, which has an eccentricity of 0, while the inclination is the angle that its orbital plane makes with the ecliptic. The axial inclination is the angle its axis of rotation makes with the ecliptic. A planet's albedo is the fraction of light that it reflects.

The number of moons of the outer planets is constantly increasing as more small moons are discovered. For the latest data, visit www.stargazing.org.uk.

PLANETARY DATA									
		Mercury	Venus	Earth	Mars	Jupiter	Saturn	Uranus	Neptune
Distance from Sun,	max.	69.7	109	152	249	816	1507	3004	4537
millions of km	mean	57.9	108.2	149.6	227.9	778	1427	2870	4497
	min.	45.9	107.4	147	206.7	741	1347	2735	4456
Orbital period		87.97d	224.7d	365.3d	687.0d	11.86y	29.46y	84.01y	164.8y
Rotation period		58.646d	243.16d	23h 56m 04s	24h 37m 23s	9h 55m 30s	10h 13m 59s	17h 14m	16h 7m
Orbital eccentricity		0.206	0.00	0.017	0.093	0.048	0.056	0.047	0.009
Orbital inclination, °		7.0	3.4	0	1.8	1.3	2.5	0.8	1.8
Axial inclination, °		2	178	23.4	24.0	3.0	26.4	98	28.8
Mass, Earth = 1		0.055	0.815	1	0.11	317.9	95.2	14.6	17.2
Volume, Earth = 1		0.056	0.86	1	0.15	1319	744	67	57
Density, water = 1		5.44	5.25	5.52	3.94	1.33	0.71	1.27	1.77
Surface gravity, Earth = 1		0.38	0.90	1	0.38	2.64	1.16	1.17	1.2
Average surface temp., °C		+167	+464	+15	−65	−110	−140	−195	−200
Albedo		0.11	0.65	0.37	0.15	0.52	0.47	0.51	0.41
Diameter, km (equatorial)		4878	12,104	12,756	6794	143,884	120,536	51,118	50,538
Maximum magnitude		−1.9	−4.4	–	−2.8	−2.6	−0.3	+5.6	+7.7
Number of moons		0	0	1	2	63	62	27	13

Stellar Data

THE TEN NEAREST STARS				
Star	Distance	Magnitude	RA h m	Dec ° '
Sun	149,600,000 km	-26.78	– –	– –
Proxima Centauri	4.22 light years	11.01	14 30	-62 41
Alpha Centauri	4.38 light years	-0.28	14 40	-60 50
Barnard's Star	5.93 light years	9.51	17 58	+04 40
Wolf 359	7.8 light years	13.5	10 57	+07 01
Lalande 21185	8.31 light years	7.51	11 03	+35 59
UV Ceti	8.54 light years	12.00	01 39	-17 57
Sirius	8.58 light years	-1.44	06 45	-16 43
Ross 154	9.66 light years	10.50	18 50	-23 50
Ross 248	10.3 light years	12.29	23 42	+44 10

THE TEN BRIGHTEST STARS				
Star	Distance	Magnitude	RA h m	Dec ° '
Sun	149,600,000 km	-26.78	– –	– –
Sirius	8.58 light years	-1.44	06 45	-16 43
Canopus	308 light years	-0.62	06 24	-52 42
Alpha Centauri	4.38 light years	-0.28	14 40	-60 50
Arcturus	36.63 light years	-0.05	14 16	+19 11
Vega	24.98 light years	0.03	18 37	+38 47
Capella	42.20 light years	0.08	05 17	+46 00
Rigel	860 light years	0.12	05 14	-08 12
Procyon	11.43 light years	0.34	07 39	+05 14
Achernar	138 light years	0.50	01 38	-57 14

Looking deeper

While this atlas is a great way to learn the sky, once you know where to look you may well need more detailed star charts, or charts of specific objects that you want to locate. The easiest way to do this is to use a computer-based mapping program, sometimes referred to as planetarium software. These programs not only show the stars in the sky for a specific location and time, but also show the planets at that moment. They usually contain stars down to magnitude 10 or 15, so are capable of producing maps of small regions of sky which you can print out and use at the telescope.

Another advantage is that you can usually add data for newly discovered comets or asteroids, if you can download the data from a suitable website. Many of the programs can be used to control a Go To telescope, after you have purchased the appropriate cables, which usually come with a copy of a suitable sky program.

Typical programs include *Stellarium*, *The Sky*, *Starry Night*, *SkyMap*, *Guide*, and *Deep Space*. Links to each of these and others can be found at www.stargazing.org.uk.

If you find that any of the links given in the atlas no longer work, please check www.stargazing.org.uk where you may find updated links. We welcome feedback about this atlas from readers, which you can provide via the website or by writing to the publisher.